FUTURE LIFE

MICHEL SALOMON

MACMILLAN PUBLISHING COMPANY
New York

FUTURE LIFE

Translated by Guy Daniels

English translation copyright © 1983 by Macmillan Publishing Company.

All rights reserved. No part of this book may be reproduced or transmitted in any form or by any means, electronic or mechanical, including photocopying, recording or by any information storage and retrieval system, without permission in writing from the Publisher.

Macmillan Publishing Company
866 Third Avenue, New York, N.Y. 10022
Collier Macmillan Canada, Inc.

Library of Congress Cataloging in Publication Data

Salomon, Michel.
    Future life.

    Translation of L'avenir de la vie.
    1. Human biology—Social aspects.
2. Civilization, Modern—1950–   . 3. Biologists—Interviews.   4. Medical scientists—Interviews.
5. Forecasting.   I. Title.
QP34.5.S2513   1983   303.4'83   83-11995
ISBN 0-02-606770-6

10 9 8 7 6 5 4 3 2 1

Printed in the United States of America

Originally published in France as *L'Avenir de la Vie,* copyright Editions Seghers, Paris, 1981.

TO *Alain Boulloche*

Contents

Acknowledgments		ix
Scientists in Search of the Future		xi
ANDRÉ COURNAND	*A Futurologist Considers His Past*	1
ROBERT GOOD	*Crusade Against Cancer*	18
ROY VAGELOS	*Drugs for the Year 2000*	35
KONRAD LORENZ	*Animals, Man, and the Patriarch of Altenberg*	52
CHRISTIAN DE DUVE	*A Great Flemish Gentleman*	73
RENÉ DUBOS	*A Bet on Man*	89
ERWIN CHARGAFF	*The Beginnings of a New Barbarism*	102
ANDRÉ LWOFF	*Passion and Reason*	120

Gabriel G. Nahas	*Pleasure and Dependence*	134
Floyd Bloom	*The Brain Connection*	150
Henri Laborit	*A Futurology of Happiness*	161
Jacques Attali	*Medicine Under Prosecution*	177
Elie Shneour	*The Deeds of Mother Nature*	190
Jonas Salk	*A Stoic of Our Time*	213
José Delgado	*Colorfulness and Exactitude*	235
Hans Krebs	*Weimar's Autumn*	249
Niko Tinbergen	*Science in a Wrecked World*	262
Jean Bernard	*A Distinguished Professor of Medicine*	280

Acknowledgments

THE AUTHOR thanks the numerous people and institutions that helped him in this task: libraries, hospital services, universities, laboratories, and other institutions. Unable to cite them all, he would particularly like to mention here: the Salk Institute, Scripps and Biosystems in San Diego, California; Stanford Research Institute and its pharmaceutical department, inspired by Mr. Von Haunalter; the University of Berkeley, also in California; the Columbia University in New York and the one in Montreal; Université Catholique de Louvain and U.L.B. in Brussels; Pharma-Information in Basel; Hoechst Pharmaceutical Laboratories in Frankfurt and Paris; Merck Sharp and Dohme in Rahway, New Jersey; and most especially the French laboratories of Roussel-Uclaf.

The author also thanks his co-workers and friends at Prospective et Santé Publique who have encouraged him and helped him in his work, and he particularly thanks Odile Robert, who had the patience to reread the long French manuscript with the attention she always brings to her work.

Scientists in Search of the Future

WE ARE LIVING in the expectation of the second millennium, the coming of the messiah or of the apocalypse, just as our forefathers lived through fear and hope of the year 1000. Therefore, the present is particularly favorable not only to prophets, diviners, and fortune-tellers—whose predictions, via newspapers, radio, have become a common social phenomenon—but also to "scientific" futurologists who try to detect, among today's burgeoning ideas, anticipations and prospects, some facts pregnant with the future.

Between the foreseeable or "futurible"—a neologism coined by Bertrand de Jouvenel in a contraction of "future" and "possible"—and a utopian ideal, the margin is often very narrow. Between the borders of the likely possible and the untramelled imaginary, there is a vast no-man's-land where few dare expose themselves. It is this field that I—together with some great minds of our time—have chosen to explore. It is the field of health in the broadest sense: man's place in the ecosystem, in life.

Yet, in my opinion, reflecting on the future seems to be a rather privileged ground for those whose research areas are life sciences, be they anthropologists, physicians, traditional biologists, or bioengineers, the last being simultaneously therapeutists and technicians who have just created a new discipline whereby man is no longer the master or the servant of the machine but is married to or integrated into it.

Of course, utopia is not the specific field of the life sciences. Social, urban, or technical utopias are more conducive to fantasy. Yet, many a utopia has become routine reality in laboratories and hospitals all over the world.

The miracle and amazement have already dissipated for some of the great conquests of health which, however, are recent. There are antibiotics, which rendered harmless some of the most virulent infectious diseases; psychotropic drugs, which have provided mental patients with gentle and often effective medication, replacing practices more akin to jail and torture than to medicine; polio and other vaccines; and more. The healing of all diseases, and the Faustian dream of eternal youth, belong to man's most deeply rooted myths, everywhere, throughout time. Can biomedical progress realize it?

For the year 2000, scientists have forecast safe organ transplants and sex determination, and for shortly thereafter—between 2000 and 2500—memory correction, prolongation of virility, creation of an artificial placenta, and life expectancy prolonged without major failings until the age of 120 years. Are these men idle dreamers? Are they illusion merchants, or insane scientists like Dr. Frankenstein, Mr. Hyde, and other characters of horror novels?

In order to maintain some homogeneity in these interviews, in which men of various cultural backgrounds in all kinds of life sciences participated, I started from a set of questions. This 20-point questionnaire—obviously restricted—is as follows:

1. From a miracle cure for cancer to a contemporary version of the elixir of youth, a whole range of medicines and methods outlines tomorrow's medicine. What, in your view, seems to be a reasonable projection?
2. What is utopia?
3. Isn't utopia particularly dangerous in the realm of health?
4. Of the medications of the future, which seem to you to be the most likely and the most promising?
5. Do you see man of the twenty-first century as being less aggressive, more convivial? Or will changes in life—brought about by demographic explosion, greater density of urban living, greater scarcity of natural resources—make him even more aggressive?
6. Does genetic engineering promise a golden age or an apocalypse?

7. Is living 120 years possible? Is it desirable?

8. Could euthanasia tomorrow, under social and political restraint, be part of a new morality?

9. Do you think that free will and liberty will be alienated by new psychotropic drugs?

10. Can we take precautions, within a democratic society, against excesses of this nature, i.e., manipulation of the psyche? How?

11. Can mental illness benefit seriously from manipulations of the psyche through either psychotropic drugs or electronic means?

12. In a world of contraceptive vaccines and test-tube babies, will conception and sexuality be totally dissociated?

13. Can you see a world where, thanks to prosthesis and grafts, failing organs will be replaced like parts of an automobile? What ethical considerations will be associated with the establishment of organ banks, which will be necessary?

14. Do you think that, for demographic reasons or because of exhaustion of natural resources, man will establish habitats in space, on the sea, in large communities?

15. Can you imagine the day when immunology will palliate the failures of chemotherapy and surgery?

16. Isn't computer-controlled public health care a prelude to an even more police-like conception of tomorrow's society, one that would offer no possible escape?

17. Will man be able to exercise biological control over his own body by the use of miniaturized devices made possible by microcomputers? Would it be desirable?

18. So many miracles are expected from the new biology that some, already, are speaking of this discipline as being able to respond adequately, not only to therapeutic problems, but also to food, energy, industrial and other needs in tomorrow's world. Do you believe this?

19. How do you see the role of the doctor and of "medical power" in tomorrow's society?

20. Can you conceive of preventive medicine without coercion?

Some of my interviewees answered all questions, while others responded only partially. Let us say that this questionnaire has served its purpose as a set of guidelines. And I found it exciting (I hope the reader will share my excitement) to compare various attitudes toward the future: some are optimistic, some deliberately gloomy and even apocalyptic. It is also interesting to note slight differences between any two opinions, although they may be close to each other.

A Pharmacopoeia for the Year 2000

Another document seemingly worth mentioning here is the table I have drawn up (and which has been abundantly used) of the "miracle drugs" of the year 2000, a list of which was submitted to my interviewees in their wisdom.

Let me be clear: this is definitely not a work of science fiction. Research into the drugs mentioned here is being carried out in laboratories now, and some are already being experimented with on animals. Research on new molecules which are efficient and active on diseases yet to be overcome is indeed the focus of discussions in the scientific world in developed countries.

A symposium held by the British Royal Society of Medicine estimates that it will take another 20 years or so to complete development and reach application of these new products. So it is today that the pharmacopoeia of the year 2000 is being elaborated.

By making a summary of the works of Bender, Blum, Strack, Ebrig, and Von Haunalter of the Stanford Research Institute (S.R.I.), using a forecast on psychotropic drugs made by Evans and Kline, and integrating assumptions made by Gabor and some others, I have arrived at this list.

Underneath its aspect of science fiction, this table—both optimistic and frightening, dealing as it does mainly with psychotropic drugs—is definitely not a gratuitous amusement at the present time. It has been compared with futuristic studies, mainly carried out by American scientists, some of them of recurrent polls using the method known as the Delphi Method.

Predictions Concerning the Discovery Dates of Medicines or Psychotropic Substances

SUBJECT	YEAR
Aggressiveness, control of	2000 or earlier
Allergies, control of most states	1990
Analysis, improvement of the capacity for	1985
Antibacterials, new generations of	1985
Anxiety and tension, control of	1988
Asthma, control of	1985–1990
Autoimmune illnesses	1990
Bacterial and viral illnesses	1985–1990
Beauty, deeper consciousness of	2000 or sooner
Cancer, cure for	1990

Caries (dental), prevention of	1990
Childhood, retarding adolescence by extending it	2000 or sooner
Contraceptives, female (safe, convenient, and inexpensive); male	1990–2000
Depression, alleviation of	1990
Edemas, control of	1990
Fear, creation of	2000 or sooner
Guilt, relief of feeling of	2000 or sooner
Hallucinations, creation of "jamais-vu" and "déjà-vu"	2000 or sooner
Hypertension, prevention of	1990
Intelligence, permanent stimulator of	1990–2000
Intoxicants, safe and with brief action	1990–2000
Learning, medications to improve	1985–2000
Maternal behavior, development or suppression of	2000 or sooner
Memory, increasing or decreasing of	2000 or sooner
Mental illness, alleviation of	1990
Mycoses, prevention of	1990
Neurological troubles, control of	1990
Nutrition, metabolism, and physical growth, mediators of	2000 or sooner
Obesity, control of	1990–2000
ŒDEMAS	
Psychobiological states, control of	1990–2000
Radiation, immunization against	1990
Relaxation and sleep, control of	1990–2000
Senescence, control of the process of	1990
Senility, control of	1990–2000
Sexual response, regulation of	2000 or sooner
Sociability, control of	2000 or sooner
Sleep, reduction of the need for	2000 or sooner
Spasm (of striated muscles), control of	1990
Time perception, reduction or extension of	2000 or sooner
Thromboses, control of	1990
Toxicomania, control of	1990

Three Steps for the Future

But let us leave pharmacology for other areas of prospective therapy. Here again, some of the scientists we have interviewed, such as Floyd

Bloom or José Delgado (who have talked about some of their experimental work and very probable assumptions to serious science writers, such as Jerrold Maxmen, Alfred Rosenfeld, Stephen Rose, and Allen Utke) will give us a chance to look quickly at the often too-ambiguous promises of tomorrow's medicine.

The eighties first: they are today and tomorrow. Before the end of the present decade, and without even mentioning data processing and minicomputers, which are already part of our life today, we will be able to find on the "health care market" some astonishing prosthetic and controlling devices to regulate vital organs. There will be a completely implantable nuclear-powered heart that can survive its patient and be re-implanted in another individual (this has already been successfully tested on animals) and an artificial pancreas automatically dispensing insulin to diabetics. There will be an electrocardiogram belt or bracelet providing information as accurate as today's heavy electrocardiogram equipment. The patient with chronic cardiovascular ailments could wear it constantly—except in the shower—and, in case of emergency, the device would emit an alarm with a range of several miles that would warn the physician or a special-care unit. In the same way, a small electronic brain stimulator would ease chronic headaches, enable paralyzed persons to regain the use of their limbs and even, in some cases, modify the behavior of aggressive mentally ill persons.

Artificial skin will decrease the death rate among severely burnt people by accelerating new skin growth and preventing infection. In the field of immunology, "smart spheres," covered with antibodies, would attack specific cancer-causing or infectious cells in the body, without harming any others. Safe and efficient gonorrhea vaccines will soon eliminate the most widespread of all venereal diseases. Contraceptive methods will range from the monthly pill for men and women to the contraceptive vaccine (successfully being tested on cows in India at present), as well as to injections of antibodies that coat the ovum and temporarily prevent fertilization. Surgical abortion—always traumatizing—will be either reduced or eliminated thanks to abortion-inducing drugs. Improvements in sperm-freezing techniques as well as in data-processing methods in sperm banks will permit the increase of numbers of donors and the selecting of sounder babies, perhaps with specific capabilities. It is estimated that in the United States in 1979, 250,000 children were conceived through artificial insemination. Apparently, sex determination can already be achieved with a 90 percent success rate.

For the nineties, the possibilities are even more surprising. They include artificial wombs in which a fetus can be kept alive until it is ready for birth; synthetic blood with the characteristics of natural blood; blood clot detection (blood clots being the cause of heart attacks or strokes); vaccines against influenza, hepatitis, and tooth decay; bone marrow transplants; prevention of some birth defects; plus eugenics techniques (and here we hope that no one will be misled). They also include tiny computers implanted in the human brain (prosthesis for brain deficiency or boosting physical or intellectual performance) and cloning of multi-cell organisms by parthenogenesis.

To Care For Twenty Billions

The year 2000 will witness potentialities even more baffling for today's minds. It is only twenty years from now—the span of one generation—and we remain within the limits allowed for prospective speculation and for possible futures. But here, perhaps, it would be wiser to avoid giving more importance to the year 2000 than it actually deserves: it is a plain round figure, a magic one, pregnant with expectancies and anxieties of another millennium. The year could also well be 2015 or 2030 if, by that time, human kind has not sunk into some apocalyptic disaster, be it nuclear or other. Then, if the Family of Man—which, according to the demographers, should stabilize at some 15 to 20 billions of individuals—is still organized into societies, scientists are forecasting other developments.

Hibernation—through intermittent slowing of the body processes—should considerably extend human life. In the same way, cryogenic techniques should theoretically lead to some kind of nondeath, by keeping cells of plants and animals in a state of suspended animation, and this for undetermined periods of time. Scientists in the Soviet Union have apparently succeeded in reviving a bacterium found in a 250-million-year-old potassium sample.

In the United States, a monkey brain has reportedly been transplanted to another animal and kept alive for several days. The possibility of keeping alive and healthy a disembodied brain by attaching it to artificial circulatory systems does not seem to be science fiction any longer.

The synthesis of plant and animal cells, creating new species out of nothing, generating hybrid monsters called man-plant-animal chimeras through a fusion of cells resulting in another genesis—all this is a potentiality, even if frightening for us today.

It is not a fantasy any more to think of the possible regeneration of parts of the human body: limb regeneration for instance, in the same way that lizards, lobsters, salamanders, and worms all regenerate parts of their bodies. In the laboratory, scientists have regenerated the limbs of frogs and induced partial regeneration in rats.

After this rapid overview of the future, is it impossible—if not sacrilegious—to imagine that man might, some day, in a laboratory, create what is called Life, out of biochemical compounds of all cells, from the most simple to the most complex?

Once again, we are not dealing here with some sort of fanciful science, some futuristic witchcraft, created through the imaginative mind of some Isaac Asimov or Van Vogt. Experiments are being carried out in America, in Europe, in Japan, in India, in China, in Israel, and elsewhere, in well-known universities and laboratories. Some twenty or thirty years ago, the production of artificial human insulin or somatostatine through genetic engineering would have appeared phantasmagorical. It is sufficient to read again today the medical literature of the time when Boyer and Stanley Cohen were reporting their first results in the technique of "chromosome cuttings" to understand that scientists themselves have only a limited capacity for forecasting, perceiving, and receiving with grace the shocks of the future, even in areas that they are familiar with.

Even if we do not go into extreme forecasts—where man plays the part of God—even if we restrict our reflection to the next ten or twenty years, even if we limit our thought to the universe whose threshold we have practically just crossed, to discoveries that seem to be already acquired, we are, however, confronted with a huge ethical dilemma. None of our laws, religions, ethics, or ideologies have made us ready for this new world. The challenge set us by these anticipations is not a scientific one (we will find a way out). It is rather something pertaining to what is called bioethics: a new morality that we have to invent—one whose premises we have not even glimpsed.

The Unforseeable

Let us return to today, and to a more sober and peaceful view of the future. Nothing is simple. Nothing is decided; no result is ever achieved for good; and the most disconcerting changes may lead us to drift toward new horizons, promised lands or mirages, unexpected success, or a painful dead end. Not all of our interviewees have equally or even similarly participated in the psychodrama of utopian anticipation. Some of them have partially rejected or severely condemned any scenario of the future, on the grounds that such an excessively long-term projection is kind of cheap journalism with a bent for sensation at any cost. In their short- and middle-term prospective views, all of them have evinced a great prudence, supported by their doubts, their uncertainties, and their fears. In a recent report, "Sciences of Life and Society," addressed to the president of the republic, three of France's most influential scientists seem to echo the views of these scientists when they warn their eminent interlocutor and "all our governing princes" against too simplistic a view of prospective science. They write as follows: ". . . what can be expected from life sciences by the end of this century? Before attempting to answer this question and to outline what seems to be this future in biology—and its end-uses in areas of social concern—we would like to emphasize the following. First of all, it is quite a dangerous exercise for scientists to have to extrapolate the future of their science, to forecast its end-uses and possible effects on tomorrow's life: they know only too well that they are certainly going to be wrong. Indeed, the main feature of research is that it is unforeseeable, particularly in the life sciences, because of their variety and complexity. All fundamental research is two-fold. One element, directly grounded on already acquired knowledge, makes it possible to predict how the latter is likely to develop in the next five to ten years. The other element, resulting from an entirely new manner of envisaging problems or elaborating questions, remains without any prediction. At the end of the forties, no planned or concerted action could suggest to any biochemist that he should join forces with geneticists and physicists in order to constitute a molecular biology. At that time, nobody could anticipate that the chemistry of heredity would be understood before that of dendrons. By their enormous efforts, the Americans have confirmed that it was indeed possible to plan for development but not for research; possible to plan for a trip to the moon but not for treatment of cancer. It is reasonable to instigate action in the areas which are apparently the most promising ones. But it is absolutely necessary to leave an important part for the unforeseeable. Nothing will be more sterile to research work than trying to shape the future on the grounds of present knowledge. At any time, one must be able, and know how, to adapt to the unforeseeable." The scientists who have agreed to talk to me have naturally taken this view.

The purpose of this book is to provide the reader not with fantasies but with projections—if not always reasonable—that have been thought over of the short- , middle- , and long-term biomedical prospects, through the analyses, deductions, assumptions, and creative imagination (supported by experience) of some of the most famous names of science today.

These texts are not interviews in the journalistic and strict sense but the product of long talks, after which those men who have already tried to envisage the future in their particular field continue to think over topics in which they have a special competence. These are (and I wanted it that way) "profiles" as well as talks, profiles of most remarkable men, often ignored by the public, and who are remarkable in more than one respect. Some of these talks—those with Robert Good, Roy Vagelos, Konrad Lorenz, Floyd Bloom, Elie Shneour, Jonas Salk, José Delgado, Hans Krebs, and Niko Tinbergen—were conducted not in French but in English. Rather than translate them from the French of *L'Avenir de la Vie* back into English, they appear here edited from the original transcripts. All the other talks have been ably translated by Guy Daniels.

The author, a physician himself, has voluntarily remained in the background, keeping his "subject" in the fore. He has chosen to listen. Listening to the other nowadays is not easy, in this world which is altogether under- and overinformed; where the essential disappears behind detail; where the important is erased by the trivial; the complex by the simplifying slogan.

Listen to these men with me.

FUTURE LIFE

ANDRÉ COURNAND
A Futurologist Considers His Past

The future . . . is preparing man for what he has never been.—PAUL VALÈRY

WINNER OF THE NOBEL PRIZE in Medicine in 1956 for his work on cardiac catheterization, professor emeritus at Columbia University, author of numerous scientific articles and essays on ethics as they concern the sciences, André Cournand discovered futurology with the team of the Centre International de Prospective (International Center for Futurology) in France in the sixties, and subsequently applied it to his teaching in New York.

André Cournand, quickly climbing the stone stairs that lead to his Parisian apartment on the seventh floor of a building on the rue du Bac, the vestige of some princely townhouse (majestic, but without an elevator); André Cournand going about his business in the bustle of Manhattan at noon, or swimming vigorously in the stream that runs through his farm, Konkapot, in Massachusetts during a long summer weekend; André Cournand, globetrotter and lecturer at once accessible and pressed for time, seemingly artless (a form of courtesy?), with a relentless curiosity and indefatigable generosity: this is my impression of Cournand, based on a few snapshot images, memories of meetings that were all too brief. When this book is published he will be eighty-nine years old.

May Providence preserve him for us until he reaches the age of Methuselah.

An anecdote. Recently, a congress of gerontologists meeting in Morocco was left without its main speaker for the opening address when the man who was supposed to speak, a distinguished Swiss writer, withdrew. The organizer of the congress asked for my help. I telephoned André Cournand, who, always very lively, took a plane from New York almost immediately.

Once the decision had been made, the organizer of the congress began having some doubts as to whether my choice was appropriate. "He's no doubt a great scientist," he said, "but does he know anything about geriatrics?"

I could not avoid retorting that André Cournand was a living example of the wildest hopes of gerontologists; that in order to prove it, all he had to do was appear, even without making a speech. In the end, he did more than that: His speech delighted his learned audience.

It is easy to see that I have more than just professional esteem for him. It is also easy to see why I wanted to lead off this collection of interviews with a man who, born at the turn of the century, has always been concerned with the future.

M.S. You were the first person—in France, at any rate—to apply futurology to the field of medicine.

A.C. That's not entirely accurate, because at first I was interested in futurology—in its methodology—without applying it particularly to medicine.

I will even admit that I was more interested in problems of education than in developing the futurological attitude as it was applied to other fields. I remember that the president of the Rockefeller Foundation—after I had sent him a book that was a veritable compendium of the articles of Gaston Berger and his friends, translated into English under the title, *Shaping the Future*—asked me if I was familiar with articles on the futurology of health and medicine.

At the time, I had had only one experience, dating from the sixties, in that field: the organization, at Columbia University's medical school,

A Futurologist Considers His Past 3

of a seminar for which students—nurses, young biologists, and even a journalist—gathered once a week. The experiment lasted six months. All of us were volunteers. We discussed problems having to do with the organization of hospitals and medical practice, and debated ethical questions in particular. Our ideas on the different types of medical practice and the general organization of the health-care system were later incorporated into the teaching done at Columbia.

M.S. And yet you straddled the Atlantic, shuttling between the United States and France, where you were a member of Gaston Berger's famous team. And, in that group, you were concerned only with general futurology?

A.C. No. I would even say that I worked very hard for several years on what has been called "the scientist's code," which of course incorporates the scientific part of medicine and the behavior of the doctor engaged in research and experimentation. . . . As a matter of fact, I published a revised version of it in a French journal in 1977. (*Prospective et Santé*, No. 3, Autumn 1977.) That particular form of futurological reflection has always fascinated me. The only futurological aim that I had then (and I still have it) was to show that the principles on which one can found a scientific ethic—intellectual integrity and objectivity, tolerance, doubt as to accepted certainties, recognition of errors, etc.—could play a role in human interactions. It is this problem that still interests me most at present, along with that of man's relation to his environment. It has its counterpart, of course, in our relation to our intellectual, spiritual, and genetic characteristics. I was trying to show, by elaborating an ethic for our time, that the principles on which scientists founded the practice of their profession and the direction of their research could be applied to human relations. I was trying to make a kind of reply to the antiscientific spirit that was manifested then by maintaining that science itself and the principles that have allowed it to become universal could play a favorable role in fostering more understanding and harmony among human beings.

M.S. What, for you, is the dividing line between futurology and utopianism?

A.C. My preferred definition of futurology is the one given by Gaston Berger in his memorable speech to the Société Méditerranéenne de Philosophie (Mediterranean Philosophical Society) in 1953: "The necessity of constructing the present in terms of the future instead of regarding it as a secretion of the past." Some people claim that anything imaginative is futuristic. That's not necessarily true. And, it is sometimes claimed that anything utopian is futuristic. That's not accurate either. An authentic futuristic study is made in two stages: a very thoroughgoing analysis of the system as it exists at a given moment; and then, flowing from that, the construction of a future that is desirable and feasible. In the course of studying the present system, one must try to discover what Pierre Massé called "facts full of the future." For example, engaging in futurology in 1900 meant discerning the importance of radioactivity, anticipating its role in the future, and then imagining what has been called "futurible" by a bold contraction: *future* images of what is deemed pos*sible*. Unlike futurology, which tries to find in the present "facts full of the future," utopianism has not the slightest root in the present. That is the essential difference. Moreover, utopian projections are not necessarily desirable, not necessarily conceived for the benefit of mankind. It is right there, I believe, that we see the thrust of the fundamental notion of futurology, which envisages possible futures for the benefit of mankind and thus has an ethical dimension that is absolutely essential.

Futurology begins with a very precise, established fact, followed by reflection on all ensuant possibilities. If one studies, point by point, the development of certain truths in biology, one can then establish a kind of politics of the probable. Futurology, a kind of long-distance strategy, supposes feedback about the present, which again modifies the version of the future fact we then entertain.

It is a dynamic system—not a static one, as in utopian projection. The latter may be something definitively established, a goal pursued, starting from a certain reality or even a "fact" of pure imagination. But it is not, I repeat, necessarily to the benefit or advantage of humanity.

M.S. The idea of constant material progress could have passed for futurology. But it turns out that it was a utopian dream.

A.C. I agree. And, through that example, the distinction between utopianism and futurology brings us into the field of philosophy rather than of semantics. The problem humankind faced was to know whether growth and development would always follow an ascending curve. History seemed to favor those who believed it, the utopians. . . . Until very recently, we thought we were living in an open system with unlimited possibilities for the future. I believe that two factors made us change the angle of our perceptions: the view of the earth from the moon, and the energy crisis. Our earth suddenly appeared to us as generally limited, at least in certain respects. The growth curve, and in particular that of technology, was logarithmic to the point where it opened up all utopias to man. To engage in futurology today is to be able to bend and correct that curve, which is now transforming itself.

The fundamental thing is to manage to control technology, whether hard or soft, in terms of the advantages that man can derive from it in another model of growth and development.

M.S. In a way, zero population growth may be a "fact big with the future. . . ."

A.C. Yes, certainly. Sooner or later we may have to create "plateaus" in order to establish more firmly and solidly what has been achieved up to now, thanks to technology.

M.S. Do you see man of the twenty-first century as less aggressive and more congenial? Or will he be made more aggressive by the changes in his life wrought by population explosion, the increased density of the urban fabric, and the scarcity of natural resources?

A.C. Fundamentally, I'm not a pessimist. But the answer to your question depends on man's future education. He is more and more aware that he lives in a group and that pressures exerted by the group influence the decisions that he makes, his behavior. . . . As Gaston Berger used to say, "Man must get used to living in a world that is in continual transformation, and learn to be happy in such a world. The future will inevitably mean densification, massification. Whether that world will be more or less aggressive, more or less congenial, depends upon us. One must learn very early to live in a group context, because everything in this world comes down to a question of interaction. Education must make this plain early on. Gaston Berger promoted an educational system that would take the child into adulthood by bringing him or her through various stages of life within the societal unit of the group—what would amount to an apprenticeship for the collective life.

M.S. And so the world of tomorrow will have no more room for anarchists and deviants?

A.C. It will be "completed." It will be more dense and, for purposes of survival, more structured and organized. On board ship, where space is more limited than on shore, passengers must obey certain rules. More than fifty years ago Paul Valéry, presiding over a graduation ceremony, made the following remark, which I quote from memory: "The essential component of education is the mind . . . preparing man to be what he has never been." Although one may not be able to foresee the future, one can prepare to face up to it.

M.S. For many people, genetics seems to be one of the keys to the future. Many people see genetic engineering as heralding a kind of Second Coming—universal and total well-being. It is claimed that genetic manipulation will provide answers to such problems as dwindling energy and food supplies, and disease. This cosmic vision of genetic manipulation may be a parallel to the cosmic vision of atomic energy—which has been considered by some as a promise of a Golden Age, the best of all possible worlds. Others, on the contrary, proclaim that both the atom and genetic engineering will cast us into the abyss.

A.C.   To begin with, I don't believe that tomorrow will bring either a Golden Age or the apocalypse. They are absolutes—utopias really. And I don't think absolutes are attainable. For that matter, it's rather amusing to realize that the pre-Socratic philosophers already had that notion. I think the absolute is necessary as a beacon, as a vision of what is possible, generating hope. Let's take, for example, the notion of absolute equality. One can wish for it and dream of it but not really conceive of it. Since each human being is unique, how could there be absolute equality? What we must do is to see how one can take advantage of that quest . . . for the benefit of humankind. Unquestionably, we are living in an age when medicine is transforming itself, where immunology and genetics (among other things) are going to play a bigger and bigger role; they are sciences that are developing rapidly, opening up new roads and exciting prospects. Man must learn to live in an unstable, totally dynamic system in which each of the variables influences the others. Genetic engineering will not escape this constant—inevitable change—of nature, and there is no reason either to be excessively afraid of it or to expect miracles of it.

M.S.   When someone is on the verge of making a scientific breakthrough—as you did with cardiac catheterization—is that person aware that he or she is doing something crucial for the future?

A.C.   I'll try to answer you as honestly as possible. There are scientists who give the impression, speaking with hindsight, that they had foreseen all the consequences of their discovery. And sometimes they do this in good faith. It's a phenomenon called cryptomnesia, the so-called secret memory of which no one knows the origins. . . . Thirty years later, one sincerely believes that the entire process leading to the discovery had been clear from the start and that everything happened in accordance with some well-defined plan.

I was not "future-oriented" right away. But it happened that in 1932 I decided to remain in the United States to do medical research full time. Actually, after six months it was suggested that I begin working with a man I respected (a man whom, for that matter, I had met by chance) on a totally new project. The work had to do with respiratory physiopathology. Our aim was to find a means of identifying, before a thoracoplasty, those individuals who might have secondary complications. Thus, we had to devise methods for studying the lungs' various functions.

Among those functions, we first became interested in the distribution of the air we breathe. We devised a method that enabled us to study the curve of the appearance and disappearance of nitrogen in the air exhaled after the respiration of pure oxygen, and to learn if the pure oxygen inhaled was uniformly distributed to displace the nitrogen. Then we studied the alveolar-capillary function; that is, the relation between the air diffused in the alveoli and the blood in the capillaries. Eventually we wound up studying pulmonary circulation—a study that, subsequently, was very rich in results.

M.S. But you already knew that the application of your method would be extended from pulmonary exploration to become a revolutionary method of cardiac exploration?

A.C. Absolutely. At the outset of our work, my colleague had placed a catheter in the auricle in order to measure the heart rate.

M.S. Is it possible, or even desirable, to live to the age of 120? I am putting that question to you, a man of some years and one of the most sprightly and creative people I know.

A.C. It's indeed possible because, as we all know, there are already people who live to be 120. But it's hard to say whether, today, it's possible to arrive at an *average* age of 120. . . . With the knowledge available to us now, the question is, is it desirable? It would be desirable under only one condition: that increased longevity not pose problems for society that would handicap the younger generations. I believe it is possible to conceive of an increase in the elderly population. But, again, is it to be desired if, after a certain period, the very old are no longer productive, if they become a burden to the younger generations? In a way it is a problem of the disposal—and I very much hesitate to use this term for human beings—of what is called "waste." Those who no longer play a role are waste matter. So is there anything to be gained by increasing their number?

M.S. That depends on the model of society we set up and, even more so, on our model of the family. . . . The elderly, provided they retain their physical and mental faculties, would have no trouble finding their place in an extended family and a tolerant society. They could even exercise a very useful social function. . . .

A.C. Can one practice a futurology of feelings? Because the problem you pose is one of sensitivity. Projecting into the future is an act of faith if it involves the realm of emotions, of subjectivity. . . . What will human sensitivity be in fifty years? If society develops in such a way that human relations lose in large part their generosity, charity, and mutual love and respect, the problem of aging will be posed in a totally different fashion. In a way this is my attempt at making a kind of *apologia* for the monogamous family of the Judeo-Christian tradition, which I hope will survive. For the moment, all my arguments are based on that wager, or hope. I am not a practicing believer and I belong to no church. I will even say that when I speak of the divine I speak of it as a concept to be surmised, of a life experience to which I don't want to attach any label or any particular credo, except perhaps my faith in creativity. . . .

M.S. Obviously, you believe in a morality for the future. . . . Euthanasia, for example. Can it become part—under social and political constraints—of a new morality?

A.C. In my opinion, euthanasia is one of those problems that it would be preferable not to discuss publicly. The media take it up and make the most serious and important matters trivial and vulgar. But, in saying that, I don't want to defend the position of the doctor who plays God and imposes his views without an exchange of opinion with the patient, the patient's family, and society in general. Although the majority of doctors have practiced euthanasia at one time or another—that is, they have cut off an intravenous drip, and perhaps in certain circumstances have seen to it that an agony was not prolonged, that a patient's death was gentle—I wonder whether such a decision, made in all good conscience, can be codified, discussed, or subjected to legislation. We are still under the influence of what happened in Germany, and we still cannot talk about such things rationally. In certain circumstances, abortions can be induced: It is better that a child who is to be born without arms or legs not be "helped" to survive. But for me it is inconceivable that a democratic system should spell out the conditions under which killing is permissible.

M.S. For many futurologists, the future of medicine, of health care, of man himself is to be found in his normalization and conditioning via an entire panoply of psychotropic drugs. We would thus have a society that was smug and very obedient, thanks to those drugs, instruments of power and manipulation by comparison with which the methods of control imagined by Huxley, Orwell, or Burgess seem strictly elementary.

A.C. Needless to say, I am completely opposed to the use of psychotropic drugs, except when used for purely therapeutic purposes—for a well-established condition or disease. In such cases, decisions are made by qualified people, either as a group or individually, who *treat* rather than experiment. But to go from that to imagining that man can be enslaved by psychotropic drugs . . . this gets us into the realm of the imaginary, because man with a capital M doesn't exist. The number of people who would have to be injected before such drugs could have any real influence on society would be considerable.

M.S. Isn't this already the case, without Big Brother? . . . Psychotropic drugs have been put on unrestricted sale and are advertised on a massive scale so that they will be used widely, all of which has resulted in the conditioning of millions and millions of people. Many people feel they can't live without an arsenal of psychotropic drugs at their disposal. Many of them live on tranquilizers, euphoriants, or sleeping pills sold freely or distributed *larga manu*.

A.C. No medication having any influence on the psyche should be distributed without a medical prescription. The doctor must exercise his or her role as a dispenser of prescriptions in a judicious manner, with great care. Of course, one can conceive of doctors, as during the time of the Nazis, who have no conscience or who are terrorized and forced into the service of a dictatorship. But in that case the problem is not the doctor or the patient but the nature of the political regime. I would hope that people would revolt against such abominable regimes, under which the doctor is degraded just as every other citizen is.

M.S. That's a very optimistic bet on democracy and human kind. So be it. But, assuming that no such disquieting political transformation of society takes place, don't you think that the mentally ill, in the clinical sense of the term, could benefit from alteration of the states of consciousness, either by psychotropic drugs or electronic means?

A.C. The problem of mental illness is primarily a neurochemical one. I believe that the future of treatment in this area lies in the neurophysiological work that is being done right now. Electronics will certainly play a role—one that I cannot yet discern very clearly—both in diagnostics and treatment. Unquestionably, "facts full of the future" will emerge in neurochemistry and neurosurgery, which are generating procedures and therapeutic approaches that are not yet completely perfected, validated, or understood. Thanks to neurobiology, which is in its infancy now, we know that the brain's structure and metabolism are extraordinarily complex. The future of treatments for mental illness lies in a better knowledge of cerebral biochemistry.

M.S. What kind of sexuality will we have tomorrow, in a world where, from the contraceptive vaccine to the test-tube baby, conception and sexuality will be totally dissociated?

A.C. Will they, really? Eros, the notion of pleasure, will last as long as man lasts and as long as we do not find chemical methods for eliminating erotic emotion from our brains. The search for and satisfaction of sexual pleasure, regardless of cycles and seasons, are specific to human beings and unique among animal species. Again, the need to reproduce is innate. Reproduction, like survival, stems from an urge that cannot be suppressed—one that is inscribed in the genetic code. Reproduction, transference, the passage of life from one to another being are fundamental processes. So the *in vitro* reproduction of humans, the test-tube baby, and cloning will never become general practice, in my opinion—at any rate, not as long as humans remain humans and do not abandon the natural employment of their free will. . . .

M.S. But isn't it possible that we're witnessing a search for a new sexual morality based on biology rather than on social utility or religious faith?

A.C. Since the beginning of time, people have been in search of a new sexual morality. But as the saying goes, the more things change, the more they remain the same. . . .

By turns, people have been either freer (or so it seemed) or more puritanical (again, we can only go by what *seems* to have been true). Fashions have varied; but feelings, frustration, and the search for love have remained the same. I don't think there has been the slightest change since my adolescence. My youth was marked in part by the search for pleasure; more precisely, for shared pleasure. It's very important to share one's desire and one's pleasure. I'd say that it's almost a manifestation of generosity that goes beyond the demonstration of any virility or femininity. The notion of birth control—of family planning—is relatively recent. It flows from the idea that procreation is most valuable in a given society if it is not a burden for oneself and others. Thus, dissociation of the sexual act and conception already exists, whether it occurs outside of or within the framework that has been codified, legalized, encouraged by the churches and society. The future will not change that.

M.S. Can you imagine a world where, thanks to prosthetic devices and transplants, defective organs will be replaced like the parts of an automobile? How shall we devise an ethic that can govern the management of "organ banks" as they become necessary?

A.C. You don't have to *imagine* such a world. We are living in it right now, in a crude way, within certain limits. I don't believe brain transplants will ever be successful, but as for the rest. . . . Organ banks have been perfected in the United States, with a system of computers that makes it possible to send organs to the four corners of the country and even of the world. I believe in organ banks. I feel it is perfectly justifiable to set aside organs that are still viable and may serve others. I don't see any difference between that and ordinary restorative surgery.

M.S. But on that point there is considerable resistance from public opinion—in France, at any rate. The Caillavet Law doesn't seem to have changed people's attitudes. For the adversaries of that law, establishing organ banks is the same thing as condoning generalized autopsy: It provides for the systematic violation of the corporeal integrity of the dead. . . .

A.C. But one must begin with education and try to gain public acceptance. In Germany, autopsies have been compulsory since the last century. Anyone who dies in a hospital will be autopsied, unless the family opposes it. Of course that poses problems of public opinion—the churches, among other organizations. But it is a policy bearing the stamp of good sense.

M.S. Either because of a demographic explosion or the scarcity of natural resources, is it reasonable to conceive of sizeable human communities in space or at sea?

A.C. Yes, I think so. But I don't believe that the population explosion is fundamental to that idea. I cannot imagine that one day we would send entire populations into space so that they could establish settlements there. What would be gained by it? It would be better to destroy superfluous individuals or prevent undesirable births than to send people off into the galaxy. And then, too, there would be problems of economics and energy that would make such a course unfeasible. It would perhaps be easier to harness solar energy from outside the earth's atmosphere. But I believe that satellite platforms manipulated from our good old planet earth would do the job just as well.

It's possible that one day we will send little groups of men to live on another planet for a certain length of time in order to do research or conduct mining explorations. And there may also be tourists in search of new diversions. But those ventures will not serve to solve demographic problems. . . .

M.S. And on—or in—the seas?

A.C. To a certain extent, yes. After all, the first human habitats were lake villages. The example of Holland proves that one can expand the stretch of land by conquering the sea. One entire section of the East River in New York has been filled in with waste material coming in from London. Buildings twenty storeys high, including Bellevue

Hospital, have been built on that landfill. I don't, however, believe we'll solve the problems of hunger and overpopulation by this kind of thing. Those problems will be solved by a set of measures running the gamut from birth control to extending agriculture so that it can feed more people to conquering deserts and mountains, etc.

M.S. Can you imagine that one day immunology will succeed where chemotherapy and surgery have failed—and that it will become the major thrust of tomorrow's medicine?

A.C. What is immunology? It is the body's response to a foreign element that tries to introduce itself into the organism. I believe that immunology will make great strides, favoring the responses of the unique "self" to the aggressions of foreign bodies. Will we find other applications for immunology, beyond the vaccine response to bacterial aggression? We can hope, but our knowledge is still limited. . . . Take cancer. Scientists know very well—although most of the public doesn't—that we are all, so to speak, healthy "carriers" of abnormal cells, cancerous cells, which are eliminated by our bodies natural defense processes. Can those processes be reinforced and stimulated? That's the big question for immunology today, and, I think, one of the most promising roads before us. But it's not the final answer, because not all illnesses or afflictions of the organism are necessarily associated with incursions by external agents. For example, pulmonary emphysema results in a transformation of the structure of the lung due to chemical assaults that are sometimes aggravated by a genetic factor. I don't believe, however, that there exists a truly preventive immunological procedure—apart from suppressing the polluting elements that are the cause of this disease. Great advances have been made in immunology. But to want to apply immunology to all diseases is unrealistic, because it supposes that all pathological processes derive from one simple and universal cause.

M.S. One idea that is "big with the future" is the management of health care by computer. Doesn't the computer printout represent both a great hope and a great danger?

A.C. Very early in my life as a doctor I wondered why each of us did not have a medical file. I was frightened by the fact that in a great many cases it was necessary to repeat again and again the job of taking a case history—something that could have been very much simplified. I believe in the value of identification, of the medical file. I believe in the importance of recourse to the computer to organize, in part at least, the life of each individual, and to supplement his or her memory. But this notion can only really be applied, for the moment, in highly industrialized countries.

All the fears one may harbor about the abuse of a technology that otherwise has many advantages are political in nature. You won't have aided in the creation of a police state just because you have chosen to store in a computer all existing information on the health of all the state's citizens. The problem is one of maintaining our democratic values in the face of whatever progress is made in technology and science. . . .

M.S.   Will man be able to exercise biological control over his own body by means of miniaturized appliances using microprocessors? Would this be desirable?

A.C.   That is a problem associated with the more general one of the role of medicine in society. Can one condone, from the moral and ethical points of view, the use of absolutely any technique that permits a person to influence his biological responses? This becomes strictly a problem of education. If you could train everyone to control his physical responses, that would be perfect. But I don't think it's possible. I'll give you a very simple example. You could go to any "service station" to have your blood pressure taken. It seems simple. But is it really? Depending on whether you take your pressure standing, sitting, lying down, in a state of fatigue, or in a rested state, you will get considerably different readings. So taking one's blood pressure is not as simple as it seems. If, with adequate education, the man in the street learns to evaluate certain problems, as the diabetic learns to control the amount of sugar in his urine, I'm all for it. But we can't rule out the possibility that all these devices might serve no purpose and remain mere gadgets.

M.S.   People hope for so many miracles from the "new biology" that some see it as capable of providing answers not only to therapeutic problems but to nutritional, energy, industrial, and other future needs. Do you agree?

A.C.   I don't believe in a panbiology that will enable us to reach a Golden Age. Biology is a flourishing science and—like particle, or high-energy, physics—has been and still is on the brink of important developments that will expand our knowledge of the physical world. Biology is the living world; and I think, as a matter of fact, that applying biology to a growing number of areas, is natural and necessary. But I doubt whether it can solve all our problems. Each solution to a problem creates another problem. I don't believe, as Hitler believed, in "final solutions." I think that biology will enable us to take great steps forward. I hope such progress will not have perverse effects—as was true in physics and the atomic peril it has engendered. There must always, at all times, be a control in place, one similar to the continually observed biological phenomenon of inhibition as a check on stimulation. In the study of the nervous system, in genetics, in many areas, inhibitive systems (as Benacerraf has clearly shown) play a considerable role—to the point where, without them, stimuli would have no effect on the development of life as characterized by their interaction.

M.S. *Fortune* magazine published an article in which biology was virtually made out to be the heavy industry of tomorrow. And today it is in the *Wall Street Journal* that we find the most details available on the practical applications of genetic engineering. But, on the other hand, while we are witnessing the enthusiasm of these businessmen for the "new frontier" of genetics, we confront considerable uneasiness on the part of the general public and even of numerous scientists. What do you think of this situation?

A.C. Both the enthusiasm and the uneasiness are legitimate. The businessmen are perhaps operating under certain delusions, because what they are calling for will not be easy. Moreover, a number of scientists interested in the problem of introducing DNA elements into microbial cells have asked themselves if what they're doing doesn't involve certain dangers. This is a result of what one might call the notion of scientific responsibility; but some of the doubts have grown out of exaggerated statements made by enemies of science. Scientists, frightened by what the communications media let loose, asked themselves: "Shouldn't we perhaps reconsider the problem before going any further?" Then there was Asilomar, and people tried to define the conditions under which research in genetic engineering could be carried out. Guidelines were imposed. But, as time has passed, they have been relaxed because the scientific community realizes that, if certain basic precautions are taken, the danger involved in this research is no greater than that associated with work on the plague or yellow fever.

But let's get back to biology as "the heavy industry of tomorrow." It is much more than that. Biology can be applied to a number of areas that involve life; it embodies the natural forward movement of recent scientific developments. Biology is the science that will make it possible, during the next fifty or one hundred years, to make considerable progress in many broad endeavors. If that progress is controlled in respect for life, we have reason for great hope. I speak not only of people's health but of the quality of their lives. Man does not live solely in interaction with his environment: within himself he has a destiny—that imponderable something we are all in search of. So I would give an affirmative answer to your question. Yes, I believe that biology will, in a great many areas, provide answers to the needs of humanity.

M.S. How do you see the role of the doctor in tomorrow's society? Some people have already consigned him to the footnotes of history. Others denounce the abuses of "medical power."

A.C. There is a lot of demagoguery and confusion on this subject. I am completely in favor of what is called in underdeveloped, or rather developing countries *officiers de santé** and very good nurses who can recognize symptoms as a result of their long medical experience. We have their equivalents at the hospital; for example, nurses examine ill or wounded people in the emergency rooms. They do the screening and know when to call the doctor.

I believe in the efficacy of these auxiliary people, but I don't think their work is sufficient to be considered complete medical practice. I don't think French or Americans would like to be treated only by nurses or *officiers de santé*.

M.S. Do you remember the polemic involving Ivan Illich [see *Medical Nemesis*] and others, which to some extent was echoed in public opinion? That dispute, originating in what has been called the "medical population explosion," was at times instigated by doctors themselves. A large number of young doctors were showing up on the labor market. Many of them had trouble finding jobs, and some were politically radicalized. It was in that context that they echoed Illich's challenging theses. Either through idealism or through idealism mixed with egoism, those doctors developed a kind of masochistic devaluation of their own role. In trying to protest against "medical power" conceived as a by-product of the "capitalist society," against the "haughtiness" of the name professors, etc. . . . they more or less threw the baby out with the bath water.

A.C. Those ideas, largely coming out of the United States, gained influence, for the most part, from the rather ambivalent feelings that people have toward doctors and medicine. Illich and his followers have dealt their blows successively against the church, medicine, and education. They are revolutionaries. They think they are, at any rate. Unquestionably, some of the judgments made by Illich are correct so far as doctors are concerned. But on the whole I think his ideas are misconceptions. It is very dangerous to destroy the confidence one has in

---

*A person who, although not possessing a medical degree, is authorized to practice certain types of medicine. (Tr.)

a doctor—not so much for the doctor's sake as for the patients who ask for his or her help. Because who could replace the doctor? That's my first comment. The second is that the situation is not all simple, as was fully demonstrated by the seminar in which we studied the problem of the relations between the doctor and society—and that was in 1965; that is, long before that issue became fashionable. We have not yet found an infallible method for discovering, among young medical students, the future Dr. Schweitzer. We do not yet know how to determine to what degree students choose medicine to make money or to serve humanity. It's as simple and as complex as that. So it's a problem of education and a problem of choice. In France, in order to maintain high standards of instruction, 80 percent of the tremendous influx of medical students is eliminated after the first year and in accordance with criteria that have nothing to do with the practice of medicine. Thus the traditional method of selection, as it is practiced now, is absurd.

M.S. What can be done on the institutional level? That's where the problem lies. Motivations can be varied: money, social status, and, ideally, the desire to devote oneself to others. The first two motivations are material and very strong. As for the third, how can it be isolated? What kind of examination or questioning makes it possible to discern the latent idealism of an individual? Must we pauperize all doctors in order to bring it out? If we decided to pay a doctor less than a manual laborer, would that facilitate solving the matter?

A.C. I think that in a profession like ours, so critical from the social standpoint and one in which certain moral and ethical criteria of behavior should play an essential role, the selection of our future doctors should be entrusted to people of experience who have proven their competence and their devotion to medicine.

M.S. You are an illustrious doctor. You have been a teacher and a recognized authority. Is it your experience of the past that prompts you to imagine new methods of screening?

A.C. But I was also a nurse. And when I was a student taking the preparatory courses in physics and chemistry, I was already spending time at the hospital to see what medicine was—to see if I really liked the profession.

Then I joined the army, where I was a combat soldier and again a nurse. I realized that anything I could do, even emptying bedpans, would help me be able to care for people. I accepted it. So the solution—and I believe that to a certain extent the Russians have found it, as proved by Solzhenitsyn's book *Cancer Ward*—is in nurses' taking the steps necessary to become doctors. I believe that, before beginning full medical practice, a person should spend a year in a hospital as a nurse. That is one of the solutions, in my opinion. It takes ten years and good hands to make a good surgeon. We could begin rather early, even before the students enter medical school, to train them to control their movements better. In hospitals, the chief of service, instead of judging students by their expressions and responses when he is making his rounds, should see how they behave with patients, because that's the essential thing. For those who want to do research, however, the problem is totally different.

M.S.   And how are the students going to live in the meantime?

A.C.   In the United States, medical students are greatly helped in their studies. They can get scholarships, state loans, federal loans, etc. Money is not really an obstacle. To conclude, I believe that the patient/doctor relationship is of essential importance regardless of the specialty, and I hope it will remain so in the year 2000 and beyond.

ROBERT GOOD
Crusade Against Cancer

AN OPTIMISTIC IMMUNOLOGIST, "Bob" Good, with the incisive writer and biologist Lewis Thomas, directs the most prestigious cancer research center in the United States, the Sloan-Kettering Center in New York City, which is simultaneously a research center and a hospital.

An optimist, and an impassioned one—these two very American traits go together. For Robert Good, biology, like space or energy, is the new frontier to conquer, a frontier with limitless promise.

M.S.   Among the wildest utopias is a cure for cancer or cancers. After false expectations, people are coming to despair, and to hope again. The question is, where are we with cancer?

R.G.   I think we've made immense strides. Certainly in my scientific lifetime we have made real strides in developing our capacity to diagnose, to understand cancer, and there are now about fourteen different cancers that were otherwise fatal diseases when I was a young pediatrician that are now curable diseases. But I look on this as just a beginning, because we are seeing—progressively, with increasing tempo—we're seeing important new discoveries about cells and about cancer. We are going to develop the ability to prevent most cancers and to treat effectively those cancers that do occur.

Now, you say there is reason to be concerned because we have had our hopes raised too high and then we get depressed and we go on with this cyclic behavior when we cannot realize this expectation. I think the important thing is that we are taught to really appreciate the steady progress that is being made and that there is an increasing tempo of discovery, because we are in what I call "the scientific revolution in medicine." This revolution has only been going on about one hundred years, it has only been in its ascendancy for about forty years, and it will have as its consequence the capacity to prevent and treat cancer. So it is not a false hope, it is a real hope, but we musn't expect it within any particular time. I think that what was wrong with the idea that we had for a conquest of cancer program was that people were thinking of it in terms of a certain restricted number of years and that raised false hopes; the idea that we would cure and prevent cancer by the time of the American Bicentennial was a false hope.

M.S. It was a political gimmick. Put an end to cancer, like conquering space.

R.G. Well, maybe. I don't think that politics were as much involved as people think now in retrospect. I really believed it was essential that we generate more support for the scientific approach to cancer, and I think that the way it was done was legitimate, but I think it should have had cautions with it and should not have raised false hopes. I think cancer is an addressable disease. I think every single scientist that I know who is working on an aspect of cancer is convinced that cancer is now a problem that can be addressed scientifically, and that was not true about fifteen years ago.

M.S. François Jacob says there is no valuable research in the field of cancer, or any valuable cure or treatment of cancer.

R.G. That's a very snobbish view. He is a magnificent man, but he is a lovely snob. There is a lot of good research in cancer and really valuable research in cancer, and there are valuable treatments for, as I said, fourteen different forms of cancer. We can cure them.

M.S. Fourteen forms out of how many?

R.G. I would say that if you talked about all the forms of cancer you would have to talk in terms of maybe one hundred fifty in this kind of generic concept. You know that every cancer is an individual event. Children used to die regularly of leukemias, but now we have

85 percent of the children surviving more than five or six years, and that is a cure. These leukemias used to kill in three to six weeks.

M.S. Would you say 10 percent of cancers can be cured?

R.G. Oh, no. I think of the cancers that come to us here at Memorial Hospital of the Sloan-Kettering Center, our cures are between 40 and 50 percent of those that we can treat first. If *we* get them first, surgery is still the main means of curing cancer, but we are curing cancers by radiation, some by chemotherapy. I was just talking about the fourteen cancers that can be cured by chemotherapy alone. There is good research in cancer. I just have to say that I think that Jacob would agree to what is being done in this institution by Boyce, which is absolutely brilliant research and by Lloyd Old. And I don't know whether he would consider Dr. Dennitt to be working on cancer, but she does. She is one of his collaborators. So there is a lot of good research going on in cancer.

M.S. What do you see in the future for the treatment of cancer?

R.G. I think that the most encouraging and the most hopeful approach to cancer is the immunological approach. That is not far enough along to talk about immunotherapeutics or anything of that sort. The immunological approach to cancer has its greatest value in analyzing cancer, in early detection of cancer, and in the ultimate development of the means of prevention of cancer and then finally in immunotherapeutics. I think all of them will be of value, so the immunological approach is the main hope in approaching cancer. But I think there may even be the possibility of immunizing against chemical agents.

M.S. Is this a conclusion you've drawn from your antigen experiments with animals?

R.G. I think that we've come a long way not only in rats and mice but in every experimental animal where we have looked for the antigens, defining antigens occurring in cancer, but in humans. We've isolated five different forms of human cancer that look to me as though we are seeing evidence of the sufficiently specific antigens that occur with the cancer so that we can use the immunological approach in therapeutic terms or in combination with chemotherapy. So it is in man as well as the animals now that have antigens on their cancers that are foreign to the host. The generation of specific immunological responses has been difficult with cancer cells, probably in part because the cancer cells can turn on negative responses of immunity rather than just positive responses. That is a target for the immunologist.

M.S. Roughly how many Americans are suffering from one form of cancer or another.

R.G. We anticipate this year that there will be perhaps somewhere near 600,000 cases of cancer in America, of which perhaps 350,000 will be fatal.

M.S. Is the number growing or declining?

R.G. It is interesting, there is one kind of cancer that is increasing in America, and that is cancer of the lung, associated with the intake of tobacco smoke, largely in the form of cigarettes. All other cancers, site cancers, are probably declining slightly. Some of them are declining very dramatically, like stomach cancer. We are not entirely sure of the reasons, but we have some good ideas about that.

M.S. Which are?

R.G. Stomach cancer began to decline in this country when we got universally good communications, that is railroads and trucking and so on. We also got universal refrigeration, and so it was a change in diet that related to the decline. About that time the use of vitamin C was promulgated for children, and I think it is a very real possibility that just the change in the intake of vitamin C—because of fresh fruit and vegetables that so characterized the American diet, and the intake of vitamin C from artificial sources—has really interfered with the development of carcinogenic substances that act on the stomach.

M.S. Does this tie in with Linus Pauling's theories?

R.G. I think that anything that has been through Pauling's computer is worth controlled study. That has not been done yet. I'm not talking about megavitamin C, even in relatively small amounts. There is an interesting observation by a man by the name of William Robert Bruce up in Toronto. He has found that there are N nitroso compounds—probably, actually, mutagens and carcinogens—present in the gastrointestinal contents of Americans and Canadians. But if he gives these normal people large amounts of vitamin C, he can prevent the formation of those N nitroso compounds. Will that prevent colon cancer? I don't know. But I think it is a real possibility, and it is being investigated. It will be inexpensive. One of the things I would like to do in some of the underdeveloped countries, where they have an extraordinarily high frequency of stomach cancer, is to just use vitamin C. Some of them have high instances of esophagal cancer. Use vitamin C and see if we can prevent the cancers there; it would be cheap.

M.S. Could you give me an idea of what the budget is and roughly how many people are involved in this war against cancer in America.

R.G. The budget in America now to fight cancer, that is with all the basic research, clinical investigations, the attempt to apply what we know, and the control studies and so on, approaches a billion dollars a year. But I would like to put that figure in perspective for you, because I think we have got lots of money for government cancer research, but I don't think by any means that it is an overabundance. I think there is good cancer research that we would like to do that is going unfunded in spite of that very large budget. When asked what they fear the most, 65 percent of Americans given all the choices in the universe say "cancer." Now that is an awful thing, to have that fear. We don't need that, and the only way to get answers that will change that is through scientific investigation, and I think that we need an army of investigators to approach all the possible directions, good investigations, of course, and the most imaginative investigations we can possibly marshal. But I think that support for cancer research is justified. Looked at another way, since the very first grant was given to Harvard University in 1922 for the study of cancer, until yesterday, the total amount that has been used by the government to fight cancer through research is about two times what it took to put Sky Lab up in the air. We can afford more.

If you see patients dying of cancer, it is a serious problem, and it is a problem that warrants our very best minds. And if bright people are not interested in cancer because very bright people aren't working there, maybe we need a little more money to attract the bright people. A little better focus on the problems would get the answers more quickly, perhaps.

M.S. I don't think it's just a question of money, at least with some, but people don't want to be losers, and to work in the field of cancer is to work on a dead end. Some may think like that.

R.G. That's right. Some may say it's better to work on fundamental molecular biology. But you know that that is working on cancer. You see, I say that Jacob is working on cancer if he's working on the basic molecular mechanisms by which the genes operate, because that is the kind of thing that is going to give us the answer to the cancer problem. Here at this institution we are approaching cancer in many ways. The place I'm placing my best bets and my best resources is on fundamental issues that may relate to the cancer problem ultimately but molecular biology, differentiation of cells, the biology of cells, the chemistry of the cells, the nature of the cell's surfaces, these are the things that we are trying to develop here so that we can really address cancer. Cancer is so close to life, life in the sense of being immortal. The cancer cells have learned to keep replicating without dying out.

M.S. Your institution is a major element in the fight against cancer in America, but there are many others. Where would you say the strenghts were of the major anticancer forces in the United States?

R.G. In three major locations: they are in that whole world of academic biology and modern revolution in biology. That is the major, major strength, wherever you find it, in institutes, in universities, or wherever. The second strength is in facilities for cancer research like this one and others that are scattered around the country, where excellent work in cancer can go on and can be fostered. There are probably fifteen of those centers. And then I think that the other major resource in the clinical investigation is hospitals, where the questions from the clinic are being placed in focus.

M.S. Can we expect something from genetic manipulation?

R.G. I'm very positive about that. I think recombinant DNA, first of all, is among the safest technologies I have ever seen. I am not the least bit worried about making pathogens and manipulating the genes of bacteria and the way that we are going to utilize them. I think this field is so exciting that we are going to see the same sort of rewards from working with recombinant DNA that we have from solid state physics. I think that it is going to be *that kind* of a major influence on industry and on products and all the rest. And I think that it will be an extraordinary resource. Now there are a lot of problems, several discoveries are still necessary but they will be worked out.

We are working with hormones that can be made in very large quantities, we will be working with recombinant technology as soon as we know the exact structure of interferon, we'll be dissecting the molecule with the idea in mind of getting genetic material so that we can put this into bacteria and have that bacteria or yeast or even human cell make interferon for us instead of having to take if from the body's own cells. I think we are looking to recombinant DNA technology as a source of vaccines for the viruses that are associated with human cancer as well as experimental cancers, and we'll be beginning within a matter of a year or two, I am sure—if not in this institution then in other institutions.

M.S. But you are working on it?

R.G. Oh yes, what worries me as a scientific administrator—I'm three things: I'm a scientist, clinician, and scientific administrator—is here at the Sloane-Kettering Institute we may not be doing enough work yet with recombinant DNA technology, that is a potent tool, one of the most potent tools that has been introduced into the field of cell biology. Part of what I call this scientific revolution in medicine is recombinant DNA technology. It is going to be a powerful tool for human good.

Let me tell you about that revolution for a minute, because I would really like for you to understand the way I feel about it. I think that the continuation of the scientific revolution in medicine can have more influence for good in the lives of people than any of the other revolutions, I mean the Industrial Revolution, the Sexual Revolution, and the Political Revolution. I think that we are just starting to see them.

M.S. Do people in general and the doctors themselves feel that this revolution is under way?

R.G. They know it is. It started back in the mid-30s where the real availability of chemotherapeutics began, and antibiotics, the immunizations, I mean all of the things that have freed us from the fear of tuberculosis, the fear of meningitis. When I was a young man, I worked on the wards. Don't tell me I wasn't thrilled when we found that we could treat meningitis and rheumatic fever and cure them rather than have all the patients die.

M.S. In France, we use the term "biological revolution" because of the recent progress in cellular biology.

R.G. No, I am talking about the scientific revolution in medicine, that's different. Now, a part of the scientific revolution in medicine will give us what I call macromolecular and cellular engineering, and this is already in the works. For example, in my own work, just by using a technique that I introduced in 1968—maps of the donors to do bone marrow transplantation—we can now cure fourteen diseases that were otherwise regularly fatal diseases. We have got lots of others that we are going to be able to cure. That is a beginning, it is just a beginning. But, with macromolecular engineering and with genetic engineering ultimately we will be able to prevent many of these diseases. So I really think that the revolution is here, and I think that when we say that we don't see it, we take too short a time table.

In the mid-40s, what about arithroblastosis fatalis? That is not an infectious disease, that is an immunological disease. That used to be a devastating disease producing hydropic babies who died as stillborns and producing fatal anemia that resulted in heart failure. Then we got exchange transfusion, that was a cumbersome high-technology answer. But we really understood that disease, we could work a little bit with the immunity system, we could prevent it completely. It is almost malpractice not to prevent it now. But there are so many diseases, I could recite almost all the diseases that were problems when I was a young man.

M.S. But these were rare diseases.

R.G. No, rheumatic fever was not a rare disease. Now we are becoming able to prevent rheumatic fever and to prevent recurrences of rheumatic fever; it was revolutionary because that was a common disease. Pneumonia killed children; it was not a rare disease. Leukemia is an uncommon disease, and it so happens that the malignancies in childhood are the most common cause of death from disease in childhood: accidents have replaced them as the most common cause of death. Now in most of the cancers in childhood we can get between 60 and 80 percent cures. And then there are burns, that's not rare. If we had a 25 percent burn in a child, that was a fatal disease; now we rarely lose a child from burns because we know how to manage electrolytes. Cholera, that is not a rare disease. Just knowing how to give the proper fluids, it is as simple as that, by mouth even. But that was part of the scientific revolution in medicine, to be able to quantify the electrolytes and know what was screwed up. I remember those diseases that I saw on the pediatric wards. There was nephritis and nephrosis, there were whole wards of children with those diseases in the teaching hospital. We never see those patients in the teaching hospital any more because the doctors can take care of them. These are all the consequences of the scientific revolution in medicine. They're not all the cellular revolution, some of them are the ionic revolution, some of them are the molecular revolution, some are the microbial revolution. We are just beginning to get the influence of the cellular revolution, and part of it is to understand the immunity system well enough so that I can take a child that is born without an immunity system and give him a few seed cells from a sibling donor matched at the majoristic compatibility system—that we didn't even know anything about when I was a young pediatrician—and the patient would be absolutely well. This is happening already; we see it all the time. You can see what we can do with just the crude tool we have now, and compare it to what we will have ten or twenty years from now.

M.S. So you feel that we are approaching a breakthrough?

R.G. It's coming all the time. It's not a breakthrough, it's just a constant effort. You can't jump to the top of the mountain; you have to climb it one step at a time, but you know you can get there, and the way you get there is by answering the questions that are raised in the clinic with the very best possible analyses and bringing the answers back to the clinic.

M.S. Were you in China recently?

R.G. Yes, I was there twice. I've made an extensive study of medicine now in Asian countries, and I really think I know something about the comparative values. I am really an admirer of what the Chinese have done in the People's Republic of China.

M.S. What did they do in cancer?

R.G. It's really interesting. A number of things have impressed me as far as what they've done in cancer. They've achieved nothing in the laboratories; they are really just beginning. They took really simple tools, for example, liver cancer in the central coastal area of China has a frequency of maybe 30 per 100,000 per annum, that is about as frequent as most of our most frequently occurring cancers in the United States. By the time it is observed by the physician, it is a fatal disease, and there is really no sense in trying treatment although they do things like plugging up the blood vessels, so maybe they will kill more cancer than they will kill liver; that's pretty crude. But what these people did was to go out and look at their population, screening millions of people and finding that alphafeta protein described by a Russian biologist was being elevated in the blood in the presence of hepatic cancer. They found that the alphafeta protein level was also elevated in the blood of patients who had chronically active hepatitis because of the proliferative diseases associated with hepatitis; alphafeta protein is produced when the primitive cells of the liver undergo division. So when cancer is undergoing division, the protein is produced in the liver, and when in this phase of the hepatitis, the protein is also delivered into the blood. The level of alphafeta protein was sometimes diagnostic; very, very high levels were seen pretty exclusively in liver cancer, but that didn't help because those were already far advanced cancer. Then, when they found slightly elevated levels and then made serial studies of the patients and found an increase of a linear growth of the levels of alphafeta protein, they could say that's cancer. The immunologist could send the surgeon in before the liver scan was even positive, and the surgeon could make a diagnosis by bimanual palpation, find the little tumor and take it out when it was still curable. The cure now with a disease that was otherwise uniformly fatal may be 40 percent of the cases of liver cancer. That is not a great accomplishment, but there is another exciting observation: they seem to have evidence that if you take out that cancer early in the disease, before it spreads the underlying disease that gives rise to the cancer (the proliferative disease of the liver, which is associated with exposure to afflatoxins in their food, gives rise to cancer) the patients don't get new cancers. Now that really rings of what we in cancer immunology call concomitant immunity, the immunity acquired while the cancer is spreading. They have raised an interesting question that can now be investigated.

But they have done things in other areas, in the south of China they have linked the cancer of the postnasal space inextricably to Epstein-Barr virus. I don't know whether that virus is the only cause of cancer in those areas, but it is a superinfection in more than 95 percent of the patients with cancer, and that might portend a vaccine. With a hepatic virus causing it, on one hand, and Epstein-Barr virus on the other, the possibility of immunization has got to be considered. And then in certain areas they found cancer of the esophagus. Just think, 263 cases per 100,000 per annum, that's nearly two times the incidence of all cancer in America. The Chinese, through extensive studies in the field and epidemiological studies, link this cancer to the soil content of molybdenum, to the vitamin A and vitamin C borderline deficiencies in those areas, and to the exposure to the nitrosomines that are generated in their grains and in their gastrointestinal tracts from eating grains high in nitrate and nitrite because of the lack of molybdenum in the soil. Now, I don't know whether they are right, but their investigations are really provocative, and they give us new ways of looking at cancer and cancer causes. They can be addressed on a community basis.

M.S. What did you learn in China that could be related to the American situation? Is there a lesson to be learned?

R.G. I think there are lots of lessons. You know, we can all learn from each other. I just don't believe that all this learning has got to go to waste. They have a lot to learn from us from the standpoint of laboratory approaches to problems and precise scientific approaches, but I think that we have a lot to learn from them in terms of means by which we can help our people by studying relatively simple things. I wish that we had information as good as they have about cancer of the esophagus, cancer of the liver, cervical cancer, to relate to all of our major epidemics of cancer. They're relating in highly practical terms to their epidemics of cancer. They have had six epidemics of cancer, and they are really beginning to understand them in terms of what the people do and what happens to the people.

M.S. In a way, preventive medicine and all these inquiries are very easy in a totalitarian society. Are they possible in a free country? Can these methods be pursued without coercion?

R.G. I wouldn't for one minute advocate a change to a totalitarian society; the excesses of any kind of authority are too well known. That doesn't mean we can't approach these problems with the idea of prevention in mind. They can order everybody to be immunized, and everyone does it. So they eliminate all of the horror diseases associated with poliomyelitis, small pox, diphtheria, tetanus, and all of those things that are preventible by immunization. We have to do it by persuasion, but you know we can do it, I mean it is possible, it has been shown that we can do it in isolated communities. It is a matter of leadership and motivation, and I think we can make requirements so that children don't attend school who are not immunized against a contagious element where they would be dangerous to the rest of the population. I think the whole world has shown its capacity to utilize good public health measures—no matter what kind of government—in the eradication of small pox. There is no difference between the free countries and the totalitarian countries where they expect the eradication of small pox.

M.S. You mean compulsive vaccination.

R.G. Well, when you understand things well enough you don't necessarily have to have everybody immunized; it's vaccination and containment. But I think preventive measures can be used in a free society, I think that they have to be presented effectively, and I think that they have to be constantly worked on, but I think it is quite possible. Look at polio: we don't have polio in this country any more; we have got a free society; we eliminated polio by herd immunity. I think we'll be testing the first cancer vaccines in a matter of a few years by immunizing against hepatitis.

M.S. There is a tie between polio and cancer, and that is the political motivation. The decision to fight polio and the facilities were made available because it was a president's sickness.

R.G. That is what people say. Certainly the national foundation March of Dimes and the people in this country really got together by voluntary means and went after polio. But there are political motivations for these things that don't have to be authoritarian.

M.S. But isn't polio and for other reasons cancer now a political concept as well as a terrible disease?

R.G. I think that that is true, I think that not only in America but elsewhere in the world. We're all in a sense overly concerned about cancer, but being overly concerned I think we will find the wherewithal scientifically to resolve that problem. I don't think it is only cancer, I think the scientific revolution will create an ability to prevent most of the things currently called disease and an ability to treat effectively those few that do persist. Now, you'll say, "What'll you die of?" All of us have known a person or two like my Uncle Mark, who died when he was 93 and was never sick in his life, at least in that life that anyone remembered anything about, and about three weeks before he died he just kind of dried up and blew away. He didn't suffer, he didn't go through cancer and he didn't go through stroke and he didn't go through heart disease.

I think everybody has it in his genetic material to live perhaps between 95 and 105 years. A lot of people say 120, I say maybe 105 from what I can read into this stuff. In mice that same potential life span is between 36 and 40 months. Now, what we have been doing in our own research is to take animals that have a very much shorter life span and to see whether or not we can address the disease of aging. I define the diseases of aging as increased susceptibility to infection through loss of immunity function; progressive increase in autoimmune diseases like arthritis, the various diseases that occur; and vascular diseases, coronary diseases, and cancers. So we have taken model systems in experimental animals for all of those diseases developed in high frequencies with aging, and we can double or triple the life span of short-lived, autoimmune-prone, cancer-prone mice: we can double their life span simply by cutting the amount of food they take in from the time of weening. They don't develop vascular disease, and they don't develop coronary disease, and they don't develop autoimmune disease, and they don't develop cancers, and they don't develop increased susceptibility to infections.

It is very interesting because their hunger may be one of the things that protects them. If you look at their activity, they are constantly active and moving around and exercising, whereas the well-fed animals are resting. It is very, very interesting. We are investigating to see whether or not there is any effect of exercise in these experimental mice. I don't think that we are going to change people's diets by studying this, but I think we are going to understand how this central nervous system interacts with the neurohumoral mechanisms, with the endocrinological mechanisms, and with the immunological mechanisms. In other words, the interaction of the major networks in the body is going to be the way we understand how to deal with these diseases.

M.S.   Let's examine some of the realities and myths of cancer. For instance, how true is the tobacco problem? There are lots of things that have been said, and they are very controversial.

R.G.   I think that the facts are really pretty clear. The intake of tobacco, especially in the form of cigarette smoke, is clearly a major factor in the development of lung cancer. Wherever people have smoked long enough, they get lung cancer. In America, maybe 80 or more percent of all of the lung cancer is from smoking. And if we quit smoking, we can get back to normal life expectancy. It has been shown that cigarette smoking causes cancer. It precipitates certain forms of heart disease and even vascular disease. It may contribute to a lot more than lung cancer. There is a possibility that it can contribute to bladder cancer and maybe several other forms of cancer as well. If you could stop smoking you could get rid of that kind of cancer. And people who don't smoke, like the Seventh Day Adventists in this country, and people who don't eat so much, especially animal fat, again like the Seventh Day Adventists, have much less cancer than the people who do. There may be other things that contribute, but at least that's something you can look at. But knowing that cigarette and tobacco smoke causes cancer hasn't changed the use of smoking, it has only changed the distribution of the use of it. You don't see doctors smoking any more. It's a terrible addiction, it's a real addiction for some of us. I quit when Sir Richard Doll's paper came out revealing that not only could tobacco cause lung cancer, as Windor and Hammond had shown, but if you quit you could get back to normal life expectancy within five years. I was a real addict; I couldn't think for thirty days, and my pulse went down to 40 and I was a miserable, terrible soul, but I quit. I've not smoked since, and I don't have any desire to smoke. There are a few people who would do that, who are highly motivated to do that, but most people are not. So, you have to use management approaches. I'm sure that the filtered cigarettes that are low in tar will be producing less lung cancer; we may see quite a reduction as a consequence of that—maybe not an elimination, but quite a reduction. But, I think much more importantly, if we can learn just what cancer is and how it occurs, we can prevent it.

M.S. But then we have the psychological factor. Even if people know smoking causes cancer, they smoke.

R.G. Sure. We know that women who are hyperobese have eight times the frequency of uterine cancer and between two and three times the frequency of breast cancer. We can't change their dietary habits, but we can learn what it's all about, and we have our model systems in animals. There's a strain of mice that has an incidence of breast cancer of 70 to 80 percent. If we reduce their calorie intake by half, they undergo nice estracycles, they reproduce, they have better-developed breasts, but they have no breast cancer. Now what the hell is that all about? You can prevent breast cancer under those circumstances. In the midst of the scientific revolution we can understand it in molecular, cellular, endocrinological, and virulogical terms, and we can learn something about it that can possibly be deliverable. I am terribly optimistic because I know that when Tannenbaum first made the observation in the early '40s that breast cancer can be reduced in high-cancer strains of mice, there was no way he had to analyze the information. There was no way of looking at the viruses: we did not know the hormones to look at; we had no knowledge of the neuroendocrinological system; we had no knowledge of immunology. When I started to work back about that same time, we knew nothing of the molecular basis of the immunity response of antibiotics. We now have full definition of the molecular base. We knew nothing of the cell; we knew nothing of the organs that were involved; we didn't even know what the thymus did. It is a completely different ballgame. Now we approach these problems with all sorts of tools with which we can make discoveries so rapidly that it is frightening, that is why I am optimistic.

M.S. Is there any question that you would like to answer with, let us say, the immunologist and cancerologist approach?

R.G. When I look at living 120 years, genetic genius is a major promise for a golden age. I don't know how much we are going to use genetic engineering, but I am sure that we'll use some. Euthanasia is, I think, probably going to be largely unnecessarily in what I visualize as the utopian future of a scientific revolution in medicine because death will be from old age. It will not be painful; it will not be associated with senility, which many think is a great problem, because that is a disease and may even be an immunological disease, but it will be preventible and treatable the way a plastic anemia is today. So I don't think that we should get into the euthanasia field, and I think that we can avoid that. I'm no expert on the psychotropic drugs, I just think that that squash up there, the brain, has a lot to do with behavior. We will learn medical approaches to control the major psychotic diseases. Neurotic diseases, I think these may be adjustment problems that have to do with the way man and his environment interact, and I think that a lot of those will be quite treatable in highly specific terms tomorrow.

M.S. Do you see any role in cancer research to be played by microprocessors and computers?

R.G. Sure I do. Let me tell you, for example, just with what I am doing today I have organized six laboratories with really good people working with the best methodologies for quantification of the immunological responses and for studying in the test tube the way the immune response works in the patients and so on. And to handle that information, to generate that information is a huge undertaking. One of my former students, who is smarter than I am, has—with himself and his son—one technician, and by using automatic processing methodologies and computer approaches and minicomputers and microprocessing he can get better answers than I can.

M.S. I was thinking about those microprocessors that we can implant and then use feedback for self-diagnosis.

R.G. Yes, certainly. I think, for example, it would be possible to use microprocessors to quantify the levels of certain markers of tumors that occurred with some frequency. And to sample from time to time to find out whether or not there was a hazard. We don't have that yet, but I can conceive of it as being available. What we can do with microprocessing and with automation and with computers is do all of this stuff so much more effectively than we are now. We'll look back on today and say, "My God, we didn't know anything at that time!" Twenty-five years from now we will have conquered most of the things I'm talking about, like disease, the way we have conquered all those things that were the real problems as a young pediatrician.

M.S. Do you think an effective immunological treatment for cancer will be possible soon.

R.G. That's an adventure that's just beginning, but that has a sort of prehistory. At the end of the nineteenth century, in 1891 and again in 1894, W. B. Coley showed that infections caused by different germs, such as streptococci, resulted in the alleviation of cancer. Then Coley failed in his experiments to treat malignant tumors with different antibacterial vaccines. Coley's experiments were forgotten. Hopes of stimulating the immunological system as a means of combating cancer were born again between 1943 and 1959, when a series of men such as Gross, Prehn, Foley, then Klein and Old, and finally Benacerraf*, discovered that within experimental cancers there were "transplanted" antigens foreign to the host organism. Halpern in France, in the sixties, confirmed these results, with the BCG later finding that the antitumoral agent was *Corynebacterium parvum.*

The practical result of the work of these pioneers was to open the path to cancer immunotherapy. A second generation is now studying in the laboratory and the clinic, and progress will be rapid from now on. In a short time we'll be able to use interferon, the agents of tumerous necrosis, and the active elements of the thymus in the prevention and treatment of cancer. The golden doorway of cancer immunotherapy is open. This doorway constitutes a fourth method of treatment, a "gentle" method after the comparative barbarities of surgery, radiation therapy, and chemotherapy. It's possible to foresee the

---

*Nobel Prize for Medicine and Physiology in 1980 jointly with J. Dausse and G. Snell.

use of a "selective" cancer immunotherapy based on active immunization, taking place gradually, as each cancer is identified in man. Carey, Takahashi, Oettgen, Shiku, and Old have already designated a specific, seriologically detactable antigen for melanoma. Immunization with malignant cells and powerful adjuvant substances could be effective against acute myeloid leukemia and lung cancer. A co-worker in this institute, Philip Levine, has advanced the probable thesis that cancer immuno-therapy will be possible if we can sufficiently stimulate the cytotoxic antibodies that eliminate the "illegitimate" antigens on malignant cells. I could cite many other equally promising works on immunotherapy. Anticancer treatment is only one aspect of an immunotherapy whose offshoots will be the therapy of the future.

M.S. For the time being, what do you foresee in science fiction becoming reality?

R.G. In the life sciences, it's difficult and often silly to give dates that are too exact to discoveries now in the stage of clinical experimentation. For the next few years, and that could be five or ten or twenty years, I'm going to risk a few predictions. Growing bacterial disease will begin to rely, in greater and greater degree, on antibacterial immunization. If chemotherapy and antibiotic therapies fail, they will be replaced by effective vaccines. We will perfect cellular engineering and within a few years we will greatly improve treatment of many inborn errors of red cell, white cell, reticular cell function through cellular engineering, for example, bone marrow transplantation. It will achieve long immunologic tolerance that will make it possible for organ transplantation to come of age. Furthermore, I don't expect these achievements to be long in coming. We will initiate and develop macromoleclular engineering to maintain and repair the immunologic cellular machinery that will have a major influence in reducing or eliminating many of the diseases of aging. We will be able to control and correct the basis of the biologic amplification systems in effect or processes. We will develop as drugs active sites of some of the molecules that function physiologically in these processes.

In following the paths blazed by immunologists like K. F. Austin, John Hadden, Robert Hamburger, Ishizaka, and Kishimoto, we create an immunopharmacology that will finally restore well being to those suffering from allergies. To a large degree, this progress rests on effective immunization. We will perfect tissue classification and compatibility. We will push the knowledge of cells and macromolecules to the point where we can cure and prevent chronic viral infections. We'll learn to use nutritive elements as a means of assuring the maintenance and correction of some immunological functions. We will learn to master the realm of local immunity, permitting us to vanquish diseases like salmonella. We shall also gain increased power to recognize and prevent a greater number of viral infections. I have the feeling that what we think exists in this area is just the tip of the iceberg. We recognize some of these infections now as "slow" viruses that produce, as was shown in the work of Oldstone, Dixon, and their colleagues, the antigens responsible for these diseases, and which cause a great number of immunological responses. I'm sure we'll discover viral agents as the causative factor of what we now call immunological or autoimmune disease. The relationship between viruses and disease is not clear in this case: to grasp it we'll have to fall back on immunogenetics and learning more about adaptation problems in the host. But these problems will be conquered, and etiological analysis will lead to mastery of disease that derives from these infections. Gradually, immunotherapy will be used and, perhaps later, immunoprophylaxis, which might, as I've said replace the terrible weapons we now use against cancer, surgery, chemotherapy, and radiation—the latter two cytotoxic and carcinogenic in themselves—because there's nothing better.

We'll discover immunotherapy and immunoprophylaxis that will triumph over leprosy, malaria, trypanosomiasis, schiastosomiasis, and mycoses, against which today's pharmacopia is of such little use. I believe that these predictions are not dreams but visions that will be realized if our research continues to be active and creative, and if we continue to criticize each other in good faith.

Above all, we must pursue our studies in immunology, applying the basic sciences to the areas with the most pressing need. Although there should always be support for basic sciences for themselves, we must also insist that they produce practical results.

Conquering cancer is neither a holy quest nor a fantasy. But a disease that inspires so much terror must be demystified. Its treatment enters into a logic which, in itself, forms an organic whole. I believe that after its two centuries of development since Jenner, immunotherapy is at the threshold of triumph. And it will lead to the ultimate secret, the key, the structure, in sum, the truth of the whole phenomenon of death.

ROY VAGELOS
Drugs for the Year 2000

THE MOST INNOVATIVE LABORATORY in America, Merck Sharp and Dohme is considered the leading pharmaceutical company in the United States. MSD's success story is a recent one, born of the technological explosion of the second world war and the postwar period. Rahway, the headquarters in New Jersey—the "pharmaceutical state"—lives by and for MSD.

The headquarters are not a factory in the shadow of a big city, but what appear to be almost a series of neat-looking college campuses, around which spread the lawns and homes of the people of this town, most of whom work for the company.

Roy Vagelos, an academic with a worldwide reputation, is to our knowledge the first doctor and biochemist appointed as director of planning and research of such a large pharmaceutical concern. That job has traditionally been reserved for organic chemists. His appointment therefore is significant as the moment when biotechnology—particularly genetics—burst upon the pharmaceuticals marketplace.

Although Greek in origin, Roy Vagelos has Anglo-Saxon smiling cordiality, humor, and imperturbability, and the elegant nonchalance and almost stereotypical American calm so dear to Graham Greene. And a very, very watchful eye.

This is a man worth $200 million—the cost of MSD's annual research.

M.S. For a scientist like you, what is reasonable for the future, and what utopia?

R.V. In principle, one could take all the present drugs, and say, what are the possible improvements? And you'd say, aren't the present drugs good ones? The answer, of course, is that they're very good, but are they perfect? Are there areas in the present drug classes that are not optimal? And of course one can go right down the line and think of potential improvements in drug classes.

But let me take a specific example. For instance, in the area of diabetes, there is of course the use of insulin, various kinds of drugs for reducing blood sugar. But we know that the long-term health of diabetics is not perfect, that people continue to have the side effects of the disease even though the blood sugar is controlled as well as possible with today's forms of insulin or other drugs that control and regulate blood sugar. And, therefore, one can take the very optimistic view, which I happen to hold, that diabetes, which is a metabolic endocrinological disorder, will have mechanisms to control the long-term problems, such as central nervous system problems, arteriosclerosis, strokes, coronary heart disease, retinitis, and retinal degeneration that is diabetic. These processes are not controlled by present drugs. It may be that the drugs aren't delivered optimally, but certainly there is a great deal of degeneration secondary to diabetes that should be within the realm of control over the next twenty years. I think there are clues already to things that can be done.

For instance, one could hope to optimize the delivery of insulin or to go to human insulin for those few patients who react badly to animal insulin. I don't consider human insulin a breakthrough of any sort. It will help a significant but small number of patients who cannot tolerate animal insulin.

But the possibility for delivery of insulin that will control blood sugar physiologically, by a mechanism that releases insulin in response to food, certainly will give a better physiological control than even the present insulin. Now, there are other hormones that are abnormal in diabetes. One of them is glucagon. The problem is that glucagon levels

Drugs for the Year 2000

are too high. We don't know if glucagon per se causes the long-term problems of diabetes, but when you have a hormone abnormality and a disease state, at least one approach is to regulate that hormone and bring it into a normal, responsive level, and then to see whether you have affected the disease process itself.

M.S.   That's possible, and not a utopian dream.

R.V.   No, it's not utopian, but what do you mean by utopia?

M.S.   What you've explained to me is that in the present pattern of pharmaceutical research, one can project certain progress, that's obvious. Some people are working only on hypotheses that have been partially confirmed. You gave me an example with diabetes. Others need a utopia, that is, a projection of the future that may be a dream, a vision.

R.V.   What is the use of a description of utopia if one can't relate it to present realities? What does that gain for you?

M.S.   Some people need this to stir up their imagination.

R.V.   Fine. What would be an example of what you consider utopian in my field?

M.S.   For instance, some serious scientists think that cancer is going to be cured before the year 2000.

R.V.   Well, one could take every disease and say that it's going to disappear. After all, what is old age? What is aging? If one does not have a specific pathological process—such as cancer or coronary heart disease or arteriosclerosis of the brain—which one tends to relate to old age, one would not age.

M.S.   Let's go back to the case of diabetes.

R.V.   I think that, diabetes being a disease in which one knows the terminal devastation, and that there are certain processes that one can measure that are abnormal, I think one can project that these processes can be normalized. And so I would say that one should be able to control the long-term ravages of diabetes within twenty years; a diabetic person would be normalized so that his kidney function, his vision, his brain function, and his heart and arterial function, which leads to gangrene in diabetics, would be no different than in a normal person without diabetes. Is it utopian or something else? I leave that to you, because diabetes is a very prevalent disease.

M.S.   If I understand you correctly, you foresee in the next twenty years the mastery of a certain number of abnormal processes which could be articulated in a way that will bring about the cure for a disease.

R.V. I wouldn't say the cure, but the control of a disease. The cure could be brought about by introducing into diabetic patients new beta cells of the pancreas, grafting of normal cells which would then react normally in those patients who lack adequate secretion of insulin. So those people would be cured, but the others would be normalized. I mean they would have a normal life span and their organs would remain normal. Now I think that's in the next twenty years.

M.S. In what field do you foresee the first "normalizations," other than for diabetes?

R.V. I think there has been enormous extension of the knowledge of mechanisms of arteriosclerosis. Arteriosclerosis of course underlies cardiovascular disease, coronary heart disease, and arterial disease of the brain. And the progress has come in the area of understanding the control of plasma cholesterol levels. Plasma cholesterol—the synthesis of cholesterol in the human body and its regulations—in the last six or eight years has been delineated to a degree that is absolutely amazing. And it's largely due to the work of Goldstein and Brown, in Dallas, where they have shown how the levels of cholesterol are regulated in humans and in animals. That knowledge has indicated very specific targets as to how to regulate plasma cholesterol, in a way that would mimic the normal physiological regulations.

Perhaps the most direct risk factor we know today about arteriosclerosis is that those people with high levels of cholesterol are more prone to arteriosclerosis. Therefore a mechanism that, without side effects, can dramatically decrease plasma cholesterol will probably have a direct impact on the process of arteriosclerosis, which then affects essentially all the aging organs. I believe that in utopia, as we have in some races, where plasma cholesterol is extraordinarily low, there will be what is essentially a lack of the aging process in the arteries: coronary heart disease will not occur. Except when there is some kind of genetic disease of the arteries, where they're malformed, but the process by which a narrowing takes place because of the deposition of fat, cholesterol and lipids will be dramatically decreased and perhaps won't occur, because it will be pharmaceutically prevented.

M.S. As the head of MSD research, what are your research priorities?

R.V. We select those major areas of disease where new possibilities have become apparent to us, and so these areas that I am discussing are very high priorities for us. So in almost any of the major areas that we'll discuss, you can assume we have major investments in research, and are pushing very hard in those directions.

Because the attack on arteriosclerosis as I see it today has to come from the major risk factor, which is cholesterol. Although one cannot say today that the evidence is in that reducing plasma cholesterol to 50 percent of what it is in this country will dramatically do the job, everything points to that as being the primary objective. But the problem with the present drugs is that they do not adequately reduce plasma cholesterol. They don't reduce it far enough, and they have side effects. There is no drug that will normalize plasma cholesterol in the majority of people, or reduce it to a point, let's say 50 percent, so that it is down dramatically, so that when you look over a five-year period, you see 50 percent reduction in plasma cholesterol and what happened to these patients. In all the studies that have been done in which the patients could tolerate being on the drugs long enough because of side effects, the drugs we have today yield a reduction of plasma cholesterol of 10 percent, 12 percent. It's so close to being insignificant that you worry that the experiment is not valid, no matter what the results are. And, in fact, the results thus far indicate nothing. No changes, no dramatic changes. Sometimes a little improvement, sometimes a little worse. And so, whether or not we do the job, I think—in the next ten years surely—the whole business will be solved. In other words there will be drugs that reduce plasma cholesterol to a very, very significant degree, and they will answer the question: Does it have anything to do with arteriosclerosis, because it could dramatically affect it, or it could not. And that would end a fifty-year question, and one could forget about regulating diet.

M.S. Is the dietary approach feasible?

R.V. Dietary approach is not really enough. There have been lots of nutritional studies that indicate you can reduce plasma cholesterol to some degree, but they don't reduce it enough. They just can't, because the body responds in a feedback manner: that is, when a human eats any food containing cholesterol, the organs that make cholesterol, stop making it. Whereas if a human or an animal stops eating any cholesterol, the liver and all the other cells then start making it to a huge degree. So they compensate for what is missing in the diet.

M.S. Are you anticipating some major breakthrough like penicillin for infectious diseases? You know, you can say, if jokingly, there are really only two good drugs in the world: aspirin and antibiotics.

R.V. Yes, I think we will have a major weapon against plasma cholesterol. I think there could be a breakthrough in diabetes, very definitely. Whether it will be along the lines of controlling the secretion of or the introduction of insulin, so that it is very much like a normal person would have, or controlling the other abnormal parameters, I think those will be real breakthroughs, because of the very important incidence of diabetes and because of the uncontrollable side effects. People with diabetes do not die of diabetes; they die of the secondary problems of diabetes.

Let's talk about infectious diseases, since you brought it up. The bacterial infectious diseases of course have been, one would say, controlled by antibiotics, and the antibiotics have the problem that, as one introduces a new antibiotic, microorganisms adapt to that antibiotic by genetically changing and being able to live and grow in the presence of that antibiotic, either by making new enzymes that break down the antibiotic or by becoming resistant in that the antibiotic doesn't get into the microorganism, so the microorganisms always have learned to get around the antibiotics. And so we have the problem of resistance. Now the newest antibiotics, and those I think that we can see in the immediate future, will prevent a rapid resistance problem. There is one area of infectious diseases that really has not been adequately utilized, where utilization is a major problem for the pharmaceutical industry, and that is vaccines. I think it is possible and surely feasible within the next twenty years to produce vaccines against the major diseases.

M.S. What about carcinogenic drugs? How can we cope with cancer if we don't know what it is about? Till now we've been talking about drugs we know the mechanisms of, more or less precisely. We know what we are doing. If it is a germ, we know the germ. We know all the processes, and we are improving drugs that are for the time being not good enough. But there are other fields, like cancer.

R.V. The drugs that we have today have such outstanding side effects that very often the patient is made rather miserable by the use of the drug. Well, what is the future for research in those fields? Clearly, the five-year cancer plan by the United States government, where it essentially gave as much money as the scientific system could digest, did not pay off. And the reason for that clearly was that we didn't have enough clues. We didn't know the causes and therefore the money went into various basic research and was enormously productive in that area. But, insofar as delivering a cure, there are clues, there are even some tangible things that are very tantalizing. I suppose the most tantalizing today is interferon. Interferon is a substance being tested on an experimental level today, where there are reports and believable reports that some of the worst forms of cancer do respond. Osteogenic sarcoma, mammary cancer, bronchogenic carcinoma, malignant melanoma; many, many forms of cancer have been reported, but very small

Drugs for the Year 2000

numbers of cases. Surely the possibility that has been raised has to be followed up, and it has to be a high priority in an organization such as this.

Now, one way to handle the problem would be to make lots of interferons, and there are some people in our society who would say that's the thing to do. The problem is there is no easy way to do that. Although one could set aside a billion dollars and build a plant to produce interferon, that interferon may not be the way a normal individual would respond himself because it would be exogenous, it would be isolated either from white blood cells, leukocytes, it would be isolated from fibroblasts grown in culture, or it would be isolated from lymphoblastoid cells grown in culture, and that last is perhaps the most productive and easiest way to produce it. On the other hand, we know that when a human is stimulated by a virus to make interferon, he makes several kinds, and we may not be introducing the right interferons for specific problems. Another approach, which perhaps is a better approach in the long run, is to have some way that would induce high levels of interferons that would be endogenous, so that the interferon made by the individual is his own. Well, that approach is very attractive . . .

M.S. But it's still in the beginning stages.

R.V. It's still in the beginning, but it's a beginning that was made by Dr. Hilleman's laboratory here a number of years ago, and I would say it's one that is not outside the realm of possibility. I think that it is a very feasible approach at this time. And therefore that **h**ypothesis, which perhaps is the most exciting today—that one can have something that counters dread cancer and those viral infections not controlled in any way today—can have a biological product that would mimic a normal response. A normal body immune mechanism. I think that that is within the realm of possibility.

M.S. What factors determine the direction of your research?

R.V. Clearly we put our emphasis where a hypothesis is either epidemiological or totally hypothetical; then we feel we can make a five- or ten-year commitment to getting an answer. It's worth the drive at that time because we are on the cutting edge of science; we have an understanding of exactly where the science is going. Science has posed significant questions, and we should be in a position to answer the questions. Now, the bottlenecks are those places beyond. If it isn't interferon, then we need a new hypothesis. We need a new point of attack and that is where we get into trouble.

M.S. I'd like your comments on this list of drug development and dates, compiled by the Stamford Research Institute.

R.V. I was amused because the preciseness of some of the dates was rather amazing. We undergo what we call strategic planning every year, where we try to predict when we are going to have made specific discoveries of product candidates. We find that the objectives change by one year, every year. Those that are thought to be on the verge of being "doable," you list for the fifth year, and then every year that year changes by one year because it has slipped. Now that does not happen with all of them, but clearly that is the problem with specific years. But I would be happy to comment.

M.S. Control of aggressiveness, for instance, is listed for the year 2000.

R.V. If one takes aggressiveness as a response to anxiety of some sort, one can certainly look within the year 2000, or earlier, perhaps even ten years earlier, to antianxiety drugs that don't have major hang-ups. The present drugs have the potential for depression, potential for addiction, those kinds of problems, and I think that an antianxiety drug without the present adverse reactions certainly is within the realm of possibility because of our understanding of the central nervous functions and mechanisms.

M.S. I think that the control of allergies should be almost next year, but I suppose that in most of the cases we can already control the allergy.

R.V. Not for serious allergies. I'm thinking of the immediate hypersensitivity reactions such as asthma. In the recent past there has been a very important breakthrough, discovery of a compound called SRSA. This was a hypothetical compound whose existence was indicated by experiments. It was thought to underlie immediate hypersensitivity reactions and asthma, causing perhaps 80 percent of asthma. Its structure was reported, and it is a product of the pathway of prostaglandin synthesis. Now with understanding of SRSA, one should be able to pharmacologically either control its biosynthesis, by blocking it, or to make an antagonist. Therefore, there is a new way to potentially prevent asthma. Asthma is not controllable today. People die of asthma; they essentially choke to death, and that is an important objective. . . . I would say by 1985, the cure would be viable.

M.S. Are you working in antibacterials or antibiotics from the sea?

R.V. No, our major sources for antibacterials are soil samples, really, and chemists. The new antibacterial agents are either products of fermentation or products of our chemists.

M.S. Some big companies as you know, are putting a lot of hope and money into research with maritime substances.

R.V. It hasn't been very productive and the experience here is that by understanding the specific physiology of microorganisms and molecular biology and biochemistry, one can invent very specific assays for the discovery of new antibacterials. The laboratory has been extraordinarily successful in coming up with new antibacterials that have entirely different spectra, a much broader spectrum of activity and much higher potency, and specifically capable of controlling microorganisms that have become resistant to the major antibacterials. By using microbiology, biochemistry, and molecular biology, they have been extraordinarily productive. So we're very optimistic that new classes of antibiotics will continue to come and will control new generations of organisms that are resistant and organisms that have never been controlled before. Surely those organisms controlled by currently used drugs that have severe side effects, such as kidney side effects and/or auditory side effects, should be available through our approaches.

M.S. What of a drug to improve the learning process?

R.V. I think the learning process will become much better understood because of the new breakthrough—I'm sure you're aware of it—the discovery of the various peptides of the central nervous system, which have been elucidated in the last ten years, and which will multiply. I believe that this is just the beginning, so that one will be able to understand the mechanisms of the various learning processes. And I think those peptides that have been already discovered will be modified, and some of these may have possibilities in the learning process.

New peptides also, which are neurotransmitters in the nervous system, will come along, which should make a contribution to the learning process. And not only early in life, but to extend the learning period of a human being, because there is a time when you start to forget faster than you learn. Perhaps this is a deficiency of certain processes you have earlier in life that can be supplemented in later years and thereby extend the learning process. So I would say that by the year 2000 one should be able to improve the learning process.

M.S. What do you see for autoimmune diseases?

R.V. Of course one of the areas of significant advance in the past ten years has been immunology, and immunology has been, other than vaccines, not an area of important significance to the pharmaceutical industry, because it was not clear where the contributions could be made. The significance of the immune response to such things as rheumatoid arthritis, lupus erythematosis, or the various autoimmune diseases is now suggesting that, as the immune response is understood, one will be able to interfere with the series of reactions termed the immune response, pharmaceutically and pharmacologically, and that something can be done. What year? Oh, I would say 1990, 2000, one should have the ability to do that.

I think the significance is that there are permanent changes of the autoimmune diseases, such as rheumatoid arthritis—the deformations that tend to go on even with optimal therapy where one can regulate the signs of inflammation and swelling. By limiting the immune response, in regulating it, the long-term problems of autoimmune disease should be controllable. I would say surely not before the year 2000.

M.S. Bacterial and viral disease: I think we could be more optimistic there.

R.V. In fact, the problem with bacterial infections is that there are some bacteria that, although controlled by antibiotics, are rapid and devastating in the two extreme age groups—the very young and the very old. When people in these age groups are attacked by something as simple as the pneumoccocus, they die because they have a devastating, immediate, overwhelming infection. They die not of pneumonia but of septicemia or meningitis. By the time they come to the hospital they are already in coma, so what is the approach there? Our approach is vaccines.

Now, I think by understanding the molecular biology of viruses, antiviral agents are going to become an important field. They have not been important; there are no really good antiviral agents. The pharmaceutical industry went through a stage of screening for antivirals a number of years ago, became discouraged, and largely dropped it. Because of molecular biology, biochemistry, and microbiology, and of viruses being better understood, laboratories are again turning their attention to the reactions that occur in viruses that are different from what happens in the host, and those reactions obviously are going to become targets for the chemists, microbiologists, and immunologists. Of course immunology is going to become very important. And the possibility of vaccines of course is a very significant one within our company and our laboratory. We have a huge dedication to this effort. One of our problems is that when we make the vaccines, they are not adequately utilized. It's a problem of educating not only the physician but people that it is worthwhile to be immunized even though the incidence of the disease doesn't affect you at the moment. Our govern-

Drugs for the Year 2000

ment talks about preventive medicine, but they don't underwrite the vaccination and immunization of the population when vaccines can be made. Now obviously there are many areas where we don't have vaccines. My fear is that, even as we have breakthroughs in the areas, they are not adequately used, and that is ridiculous. To have a vaccine sitting on the shelf, and have people dying in this country and in the world. But we certainly have that situation.

M.S. The end of cancer by 1990—that's too optimistic.

R.V. Yes, 1990 is certainly too optimistic. I think by 1990 we will have solved the question of interferon, that we will have enough interferons available. I am hopeful that we will have interferon inducers adequate to answer the question whether that is a major way to handle cancer. Perhaps by the year 2000.

M.S. I think we are going to have understanding of a disease's mechanism and the drugs together.

R.V. They go together, absolutely. It happened for poliomyelitis, for instance, where the knowledge of the sickness and the immunology treatment went together, instead of having first the knowledge of the disease and then discovering the drug. But that is not the case now. If they have to go together, then we are in for a long siege. There are instances where the disease process can be ameliorated without knowledge of the mechanism, and I have been certainly brought into contact with that knowledge, since I am in the pharmaceutical industry. Let me give you an example. Aspirin was available long before an understanding of prostaglandins. Morphine was available long before the opioid receptor was understood. Indomethacin was discovered long before its mechanism at the level of prostaglandin biosynthesis—that came long after—and also the availability of major tranquilizers for the treatment and amelioration of the major psychoses. All these happened long before any understanding of their mechanism of action. So one can hope, but they're long shots, and we'll use the more traditional step-by-step understanding.

But one can hope that there will be accidental, serendipitous breakthroughs as well, but they are unpredictable. Entirely unpredictable, just as the major tranquilizers were. And in a pharmaceutical industry such as Merck, a company laboratory such as this, we do the basic science and we stay on top of all basic science, that is our major investment. On the other hand, when there are hypotheses that come along, that can be tested, even though they are not adequately based in basic science, we sometimes feel that we have to test them, even though the science isn't there. Now what is the balance between those two? Our great emphasis is basic research, and what is predictable from where the science is. By all odds, that is our major emphasis.

M.S. No more cavities by 1985?

R.V. Dental caries. Well there's only one thing that I would raise in this regard, it's been discussed, and that is a vaccine against an organism thought to be causative, and that is streptococcus mutants. Such a vaccine has been attempted for a number of years; there is an experimental model available in many places. One can test such vaccines, and studies are being done in a number of places, including in our laboratory. The possibility of testing that hypothesis I say is surely within 1990. And that could be a very significant breakthrough in health, because you know what bad teeth can mean for the health of an individual.

M.S. Let's go to hypertension. Cured by 1990 or 2000?

R.V. Well, the reason for those years, I'm sure is that there was an announcement by another company, Squibb, of a specific target, an enzyme known for a number of years, that converts angiotension I to angiotension II. I have to assume that that gave rise to the dates. That mechanism as a target for regulation of blood pressure is a very reasonable one, and what Squibb scientists did was make a good inhibitor that's been shown to work in the clinic. It does reduce high blood pressure in some patients. That drug is not optimal, but the mechanism is a new one, and therefore it will be utilized. It is not a panacea for control of high blood pressure. What I see in hypertension is a disease with multiple causes, and therefore there will be multiple mechanisms for attack, for regulation. The control of any individual patient will still be something that has to be determined with that individual, and which drug and which mechanism is functioning to cause the hypertension will be very specific. Therefore a number of drugs will be utilized, until all the causes of hypertension are understood. If one can target them on specific causes in individuals and understand in person A, B, or C, what the cause of his hypertension is, let's use the drug that affects that mechanism. That's the future of the treatment of hypertension. You know that the first really major attack on hypertension that essentially revolutionized the treatment came from this laboratory.

M.S. Do you think memory prolongation will be achieved in less than twenty years?

R.V. Memory is interesting because becoming less efficient with age certainly has to be a problem of arteriosclerosis. On the other hand, there must be connections within the central nervous system, which consist of certain cells and receptors on the cells and transmitters, and if aging in part is a decrease in the transmitter substance, and we understand the various transmitters, then one can hope to approach the memory problem by supplementing the natural transmitters, which are being underproduced. For instance, we know that Parkinson's Disease is a problem with dopamine, and if introduction of dopamine, by one mechanism or another, improves a central nervous system deficiency and a deficiency disease, then one could hope that many of the problems of aging and of memory are similar deficiency diseases and could be supplemented exogenously by pharmaceutical agents or drugs. I think that in the long term, by the year 2000, memory will be something that is amenable to pharmaceutical agents.

M.S. It's funny that we can't find a cure for fungal diseases. The most common mycoses, athlete's foot, for example, are just as persistent now as in the past.

R.V. They're not life-threatening, and therefore they have not been major targets. Now there are some fungal diseases that are life-threatening. In the quest for antimicrobial products with modern knowledge of the organisms, just as we can do more for antibacterials, we can do more for antimycotic agents. And I think they will be targets for innovation in the future, because they are more attractive. I think drug discovery goes in cycles where new scientific knowledge allows certain hypotheses as attacks in specific areas. Those attacks are made—they either work or they don't work. And then you need a new idea. And that new idea takes some time to develop and poses new things to attack.

M.S. How about control of neurological problems.

R.V. We talked about Parkinson's disease. I think there can be improvements to the treatment of Parkinson's Disease, because the disease in some places will continue to evolve and to progress even with the best agents, so drugs that have new mechanisms of action, I would assume, will be coming in ten, twenty years. Other neurological diseases, certainly congenital diseases, the spastic diseases, degenerative diseases as amyloidosis, epilepsy, multiple sclerosis, I think are amenable to testing hypotheses now. I would think neurological disease is high on the list, we'd be able to do something in the next ten or twenty years.

M.S. And edemas?

R.V. Edemas, of course, I think are controllable: that's water and salt. There are new generations of diuretics in the laboratories even now and I suppose one could look at a diuretic that gets rid of water and salt and does not cause any side effect. That is not available today, but certainly by 1990. I would say a good possibility would be that diuretics that do not cause a retention of potassium, uric acid, or elevation of blood glucose—the side effects that are rather limiting and cause worries today—will be achievable in the next ten, twenty years.

M.S. Do you think there will be any way to immunize against radiation?

R.V. Well, if you ask biologists, they will say that you will in time mutate, because the biological system is understood. If you take organisms and expose them to high levels of radiation, you select those organisms that are resistant, and you can select by going through many, many generations, certainly in bacteria, rather radiation-resistant organisms. Now those organisms have a way to repair damage. One mechanism is to select for organisms that repair the damage done by the radiation in their DNA, so they have evolved enzymes that are DNA-repair enzymes. Now, to do that in humans of course is not in the immediate future. To have drugs which in some way block the effects is not unthinkable, but that's utopia. It's not clear how we would do that. Radiation does several things. The major thing it does that we fear is damage to DNA, and we don't know how to prevent damage to DNA.

M.S. The regulation of sexual response is projected for the year 2000. Will this be a question addressed by pharmacologists?

R.V. I wouldn't put this in the hands of psychiatrists because I believe most psychiatric diseases are really biochemical problems. There are some very interesting studies, for instance, the changes in hormone levels in the brain, in relationship to sexual activities, are very exciting from that point of view. One could, in time, hope to understand what sexual behavior is controlled by chemical changes in the brain and affect us. If homosexuality is based on a biochemical difference between heterosexual and the homosexual, that should be amenable to pharmacological action.

M.S. What can be hoped for in the area of sleep control?

R.V. I think mechanisms and drugs that will give normal sleep are attainable, but not before 1990. One must understand the mechanism of normal sleep. The present drugs cause hangover. They still cause some level of depression, of reaction, and they're not normal—those are not aftereffects of normal sleep.

M.S. What do you foresee in the industrial application of genetic engineering?

R.V. In regards to the use of the present understanding of genes and change of genetic material in microorganisms, I think that will certainly give rise to mechanisms for obtaining peptides and other compounds from bacteria, so that genetic engineering will definitely have a very productive use in the pharmaceutical industry. The major objectives of genetic engineering within microorganisms surely will be the synthesis of small peptides and large peptides. Insulin, growth hormone, somatostatin are peptides that can be produced. How important this will be will depend on the ability to synthesize large amounts of these peptides chemically versus microbiologically, the economics of the situation, and the medical significance of the peptides. Making such things as antibiotics, other molecules that are not peptides, by genetic engineering, that's going to take longer, perhaps another ten years, perhaps twenty years, before one can have small factories of microorganisms that produce large amounts of unusual compounds.

Genetic engineering as it pertains to humans? I think that is much further down the line. Selection of clones of people, who will be engineered by selection; who should reproduce and who won't reproduce: again, political problems. How about genetic engineering for people who have genetic deficiency diseases? That is the introduction of normal DNA into people who have defective DNA by congenital problems, or genetic problems. That is something which we don't have. I would say it's not in the immediate future, but twenty years from now, perhaps the introduction of exogenous DNA that would repair a deficiency: phenylketonuria in people who lack a specific enzyme or have a deficiency in an enzyme; the lipidoses, the lipid-storage diseases that have genetic deficiency in an enzyme that normally breaks down lipid in the brain or in the body, and therefore you have overaccumulations of lipids, which causes a disease state. A congenital disease, then.

It's hard to see that we're going to accomplish that in the immediate future because we don't know how to introduce normal DNA and get to the right target, get to the nucleus of the cell, where you'll have it used as endogenous DNA to make messenger RNA, which is translated normally. So the isolation of normal DNA and getting it into a human is still not within our capability. But in utopia, it certainly is because we're doing it with bacteria.

M.S. Can man accept the alienation of his freedom of choice, his liberty, by psychotropic drugs?

R.V. As we develop drugs for psychiatric use, psychiatric disease, the best drugs should be those that allow a person to function normally and to make normal choices, and those that leave a person in a glazed, vegetable-like condition are not going to be very popular, because especially in this country, people are very motivated to having freedom of choice, and drugs that don't allow for that will not be used and will not be allowed. So I would side very definitely with psychotropic medicines based on biochemical understanding of the nervous system, as by far the most productive areas for the future.

M.S. Will the introduction of contraceptive vaccines dissociate sexuality from reproduction?

R.V. I believe contraceptive vaccines are doable; they certainly are scientifically within the realm of possibility. Again, the mores of the society are going to be very important in what is acceptable, because people want to be able to have babies when they want them, and when they do not want them, they want to have drugs or vaccines that have absolutely no side effects.

M.S. Have we entered the era of prostheses?

R.V. I think we have had an absolute revolution since you and I were medical students. I am sure you have had the occasion to see someone who is completely deformed by arthritis or from a congenital malformation of the hip who has been just racked with pain and crippled, and then suddenly he is given new joints, new knees, or new hips, and he is a completely new person. That is an absolute revolution. And I expect that this will surely increase. As the immune response is understood and controlled, that organ transplant will become much more common and acceptable.

M.S. Poliomyelitis had a major political impact in America because it was the sickness of the president of United States. If Salk had not created a vaccine, the cure of poliomyelitis today would have been very sophisticated. So a very simple immunological act made a transformation that made useless all the other technologies of medicine. Could you foresee this panimmunological way of medicine?

R.V. Well, as I said, prostheses are an example. One of the major needs is in the people who have end-stage rheumatoid arthritis, who are badly crippled and who need prostheses of various sorts: knees, hips, and joints. An immunological attack on that disease, where the autoimmune response that causes the disease can be eliminated or blocked, would be a specific example of that sort exactly.

M.S. One of the scenarios for the future includes controlling public health by computerizing all disease information. Wouldn't this lead to abuse? A more police-like state?

R.V. We have computers used in health diagnosis, so that much information is available concerning individuals, which could lead to some control. I don't see that a police state is going to happen, because of the great drive towards individualism and freedom in this country. And computers therefore will be controlled, even though they're control items. What's more, computers are being used in diagnosis of disease more and more, but they are not a panacea. They can achieve rapidly what is understood, and certainly X-ray diagnosis and the ability to localize disease processes has improved enormously by the use of computers. But I don't see any of that being used in a police-like problem in controlling society. I just don't see that as a problem in this country or anywhere.

M.S. One expects so many miracles from biology and some are already speaking of this discipline as being able to give answers not only to therapy, but also to food, energy, and industrial and other needs in tomorrow's world. Do you believe this?

R.V. The new biology is being assumed by the pharmaceutical industry as the basis for an understanding of disease of the future. Very definitely. And thereby setting up new hypotheses that will lead to new forms of therapy. And I think we have had, in the last twenty-five years, an explosion in biology, which has had remarkable results not only in treatment but in agriculture, as well. But the new biology is not going to answer all our problems, particularly not of energy.

M.S. For several centuries, science has been first by magic and religion, and then by physics. Now we are entering a third era, the era of biology.

R.V. Oh, absolutely. We are strong believers in that and we are building up for that. You know the laboratory in this corporation is building very strongly to enlarge our biological research capability, and we believe in it enough so that we're investing our money in it. And we're not going to lose any money on this bet on the future.

KONRAD LORENZ
Animals, Man, and the Patriarch of Altenberg

FIRST COME THE DOGS—two, four, half a dozen—who welcome the visitor at the gate of a somewhat ramshackle mansion, standing lopsided in a large park. It is Altenberg, one hour by train from Vienna, a small village at the Danube riverside. Heavy chest and white flowing beard; smiling and apparently very pleased with the noisy affection of his dogs, here comes Konrad Lorenz, the lord of the manor.

Later on, the Altenberg patriarch will show me his fish, which he is now studying in a huge aquarium worthy of Disneyland. To my regret, the famous greylag geese are not there. They have been penned some 50 miles away, on a Danubian island. After a careful search, one might eventually find some tortoises, snakes, and other impedimenta belonging to Lorenz's wonderland; he is an ethologist, a poet, a physician and a philosopher, and perhaps something of a wizard.

Awarded the Nobel Prize for Medicine and Physiology in 1973, he may be considered, together with Nikolaas Tinbergen—with whom he shared the award—as the founding father of ethology, the new science that investigates animal behavior and tries to isolate from it information regarding the behavior of the most disturbing animal: *Homo sapiens*.

The flamboyant works of Konrad Lorenz, his long life amidst European turmoil, and the richness of his thinking make him an essential witness of our times.

M.S. Between a miracle cure for cancer and the Fountain of Youth, the public has a very optimistic idea about the possibilities of the medicine of tomorrow.

K.L. Funnily enough, the cure for cancer and the cure for old age have something in common. I know very little about immunology, so I can only tell you what my authorities on immunology tell me. And my authorities are McFarvey Burnett—Nobel Prize winner, Australian immunologist and very, very famous man—Otto Reshfal and also Harry Fisher. They think that antibodies were "invented" in organisms of long-lived species in which the danger existed of any little mutation's producing asocial elements, i.e., neoplasms; the whole antibody apparatus is there to suppress neoplasms. And this might, of course, give some hope (and this is almost a well-supported theory) for a cure for old age; old age too is a progressive illness with an absolutely mortal prognosis.

M.S. But the aim of my question was to inquire about your thoughts as to what is reasonably possible in the future and what belongs to the realm of utopia.

K.L. That can be realized by mastering, controlling antibodies. This thing about cancer, the cure for cancer, is not so much of a utopia. There is a quite well proven theory that antibodies exist that can cope with cancer. Now, some people think that antibodies cause old age; that man, or any long-lived organism, in the course of very long life, becomes allergic to himself, destroys his own tissues by antibodies, and that aging is caused by antibodies. And this has been actually experimentally proven in some fish. If you keep them at a very low temperature, where they demonstratively cannot produce antibodies, then they will live forever. So, it is not so utopian to think that maybe, by knowing much more about antibodies, human life could be prolonged almost indefinitely. And, in a way, this would be highly desirable; indeed, if you were as old as I am, you would regret that the accumulated knowledge of old age never comes together with the creativity of youth. And if, with the knowledge I have today, I still had the creative drive I had at twenty-five, I should be a very, very great man. You see that this utopia of prolonged human life has its roots in a highly desirable thought: our wish to extend life's duration within a certain scientific reality. This is not really utopian, at least within reasonable limits of prolonged life.

M.S. Since you have studied animal aggressiveness so extensively, what do you think about the possibility of successfully controling human aggressiveness before the year 2000? Will we be in a position to overcome family conflicts, wars, and so on? The year 2000 is tomorrow and, in a way, already today.

K.L. I think that I have all the facts to understand aggression and, on the other hand, it would be silly to deny its existence because we have to cope with it. You always have to know the horse you're riding. You have to know when the horse is getting out of control, and I think that quite certainly we should be able to control the bad effects of aggression. But aggressiveness is something absolutely necessary; without aggressiveness we have no personality, we are nothing. But as regards the bad effects of aggression, I fear I haven't done much to control it before the year 2000. If people read what I've written about militant enthusiasm and understand it, they would never be caught up by it. I advice anybody who wants to know about militant enthusiasm to take one look at the military parade in Edinburgh. There you have all the stimuli—that cleverly contrived "super lord" stimuli—to arouse militant enthusiasm. And if you know that, you can watch yourself. And take your pulse when that wave of emotion comes over you, and then I think you will be more or less immune against being taken in by any demagogue ever again.

M.S. You know quite a bit about it, don't you? And what you experienced as "militant enthusiasm" happened to be nearer than Edinburgh.

K.L. Yes, unfortunately.

M.S. Do you envision substances that could control groups and have mass effects on people?

K.L. I don't think so. I believe instead in education, in teaching people the motivations for their behavior. Chemical substances against aggression are nonsense because they would destroy everything; they would eliminate all negative and positive compulsions at the same time. With the exception of drugs to combat allergies, I would say all drugs are highly dangerous because they upset an organism's equilibrium. You can never say something is simply good or bad.

M.S. I am thinking of the use of those substances, for instance, to relieve guilt.

K.L. That would be very dangerous. This would eventually end up in the disappearance of the last safeguards that protect some of us from sadism, homicidal tendencies, etc. The use of such substances would be insane.

M.S. What about the aggressiveness of people on the highway? Is that aggression?

K.L. Yes. A German journalist, Paul Klemen, has written a very interesting book about aggression in drivers. And this is more or less agrees with my most recent observations on fish, which show that the most important factor in antagonist control of aggression is personal acquaintance; being anonymous makes aggressiveness much easier. If you know your adversary, just know his name, you find it more difficult to attack him. It puts a certain control on the output of aggression. And in a car, the dangerous thing is that you don't know the other drivers; they aren't even people: you see only Renaults or Citroëns, and yet they arouse all your aggressiveness. And that's why drivers are so incredibly brutal. The worst side of human nature comes out in driving. People have been known to get out of their cars to fight. This is mainly, I think, the unbalancing effect of putting a man in a box so that he is not a man anymore, but just a machine, something brutal. But in conclusion, I would say that psychotropic substances, chemical or otherwise, would be very dangerous, except in very small things: for instance, a substance to prevent tooth decay is thinkable.

M.S. You make no exceptions? I was thinking of those psychotropic drugs against anxiety, pressure, stress, those drugs which would increase the awareness of pleasure, for instance.

K.L. I would consider these drugs dangerous because they upset the natural balance; drugs that give pleasure or eliminate fear would be very, very bad. Suppose a man driving a sports car has fear suppressed; my goodness, think of the consequences! And you know, it's a very interesting question today whether courage is still to be considered a virtue. Courage is not required of a normal civilized, city dweller. A person needn't be courageous, and courage is only useful in certain limited cases—to face down cowards. Because cowards are apt to get aggressive; they're afraid, so they hit out before they're attacked. This behavior is well known in so-called biting dogs that bite when they're afraid; and this behavior is one modern human beings still have. Otherwise courage is highly dangerous. It destroys hundreds of nice, valuable young men every year in mountain climbing. And in driving it is still more dangerous. So courage is only useful as an antidote to fear.

Yes, I'm strongly opposed to all psychotropic drugs. Generally, humanity would be better off without them. Suppose I had a pill to control obesity. The consequence would be that I would eat twice as much as I do. Well, all these are still supposed to be psychotropic drugs.

M.S. How do you see man of the twenty-first century? More aggressive or more congenial?

K.L. I don't know whether man will become more or less aggressive: less aggressive thanks to a more harmonious organization of society at large; or more aggressive because of what you call demographic explosion. Our studies on fish show that the optimum aggression is directly related to crowding; and if you increase the crowding even more, aggressiveness paradoxically decreases again. If the fish are not crowded you have minimal aggressiveness, then at a certain state of crowding it reaches a maximum, which decreases again if you crowd the fish still more. So it is not certain whether greater density will harm humanity by creating greater aggressivity, but it would quite certainly cause other damage. The demographic explosion, which implies a technological explosion, would mean that everyone has a narrower and narrower field of knowledge in which he must be absolutely expert in order to compete with other people. It reminds me of the old joke about the doctor, the specialist who knows more and more about less and less and finally knows everything about nothing, and the general practitioner who knows less and less about more and more until finally he knows nothing about everything. That is perfectly true, and we become more and more dependent on experts just as we become more expert ourselves. The small slice of expertise which a man can master fills up his time in his life, in his learning capacity to such an extent that now the man has no time anymore to know about anything else. And that in my opinion is one of the greatest dangers of the urbanization of demographic explosion. Aggressiveness is relatively harmless compared with the other narrowing of human mentality.

M.S. You mean everyone will have to be satisfied with fitting into a shoebox and feeling all right intellectually and physically.

K.L. Exactly. More than aggressiveness, I fear bureaucracy, compartmentalization, overspecialization, robotizing, etc.

M.S. I would be interested to know more about your fish. You were mentioning two stages: one of nonaggression and . . .

K.L. Not nonaggression but a curve of aggression, density-intensity increasing on the abcissa, aggression represented on the ordinate, gives a curve like that.

M.S. Then when and why are they less aggressive when they are more crowded, because then they have to find room for everybody?

K.L. No, they give up having room for themselves, and I think that men do the same. I think that the Chinese who are crowded very very much are less aggressive than you and I are. That is possible.

And physiologically I think that aggression is present all the time; internal production is constant, and there is always just a little bit of aggression because there is always another individual too near and in that way aggression sort of seeps out and cannot reach higher levels because it is always drawn away. It sort of sinks.

M.S. Well, in the Wild West they had a lot of space.

K.L. Yes. So there is an optimum crowding for aggression: there is one degree of crowding where aggression is greatest.

M.S. Did you find it in your studies of fish? Did you find it in the so-called notion of territory among them?

K.L. Yes, it is everywhere. But look, it is very difficult to do that kind of experiment with birds or with mammals, it is too expensive, but with fish you can do it in an aquarium. But it is quite certain that we see the same phenomena in birds and mammals too, quite certainly.

We can see it in a zoo with mammals, if we increase or decrease the number of animals. These things have been done in mammals with a much higher social organization and there, of course, different laws pervail. Ackman has done highly interesting experiments on crowding and aggression with rhesus monkeys, and we know from field observations that it is very much the same with baboons. In these monkeys, the highest ranking ones, the rulers of society, are not single individuals but coteries—generally two or more very old males who are good friends—and they support each other. They are rulers of long standing and keep their supremacy by being friends and by sticking together.

M.S. Like mafiosi?

K.L. Something like mafiosi or like the lobbies of the great industries. They are very similar. A collegium.

Now this collegium is very beneficial to the monkeys, because the leaders are the oldest, most experienced, and most intelligent of the society. Now, if you crowd them, everybody gets nervous and everybody gets more aggressive. And while these three or four rulers are very benevolent, kind old gentlemen, when you crowd them, they begin to fall out, they begin to fight with each other and lose their coherence as a group, and, from that moment on, they are dethroned by strong young males who then rule supreme, who are not so intelligent but rule by brute force. And the ethologists found an amusing thing in the two types of monkeys, in baboons and in rhesus monkeys: you can measure aggression very simply by counting how often one frowns at his friend. He doesn't do much more than that, and if you measure aggression that way in these senates—which of course had a small ranking order—there the president and the vice president are good friends and there is a minimum of aggression measured between these three or four, practically no aggression between number one and number two, they are just friends. And then aggression in the middle ranks is greater and drops down again in the very low ranks. All the animals that are beaten by everybody are very unaggressive toward each other. Now, if you change that background and you have a single tyrant, this really is a picture of democracy against a totalitarian regime. If you crowd them, increasing aggression, then you get a totalitarian ruler; you get the Hitler-Mussolini pair, and you find maximum aggression between number one and number two. And I think that is highly interesting. And that is very beautiful, I think. Very, very worthy of thought.

M.S. What do you think about genetic engineering? It's the big thing today, the biggest threat and the greatest promise, just like the atom.

K.L. Well, I think it's quite as dangerous to touch that as nuclear power. At the present state of our knowledge, we do not know anything like enough to tamper with our genetic code.

What is our notion of ideal man? You and I aren't behaviorists. B. F. Skinner believes that you needn't tamper with genetic codes because you can teach a man to be the ideal non-autonomous man. Behaviorists and genetic tinkerers are all megalomaniacs. Who would dare; it is hubris; it is a sin to believe that you can play God. It is playing God and we just don't know enough. We can begin very gingerly by eliminating hereditary illnesses; it's a very good thing that we can examine the genoma of the amnion and prevent a Mongoloid from being born into a family. My answer to the threat of genetic engineering is that the Golden Age may come in the year of our lord 6000; at the present time the promise is of an apocalypse.

M.S. Well, the problem is that you cannot put the genie back into the bottle. It's out. Either we find a way to have a code of behavior—a way to control genetic tampering—or we have disaster, but we cannot put the genie back into the bottle.

K.L. Well, nobody would really dare it. All engineering, "social engineering" is a phrase I abhor. It is very simple. Genetic engineering, according to Dr. Skinner and similar people, is not necessary because man is born as an empty organism and can be manipulated any way you want. And if that theory were true you would not need genetic engineering. Now, many responses of man are merely genetically fixed behavior patterns: wanting to have a house, wanting to have a wife, wanting to have a child, wanting to have a garden. The simplest elementary human rights are fortunately fixed and if we open the door to a diabolical manipulator, if we permit him to breed men who do not object to being manipulated, then human freedom would be in the can and that is what the mafiosi actually want. It is no surprise except to me that religion is used to manipulate man the same way among the big American producers and among the totalitarian Bolshevists. Because their aim, their ideal, is to have a man who can be unlimitedly manipulated to be an ideal consumer, to buy what I produce or to be the ideal obedient Soviet soldier. And the answer to this is also *quis-custodiet*, who shall watch the watchers, who shall control the controllers. And the mere idea of Dr. Skinner having a say in what shall be done genetically with man . . . I don't believe in a personal God but that shall be the victory of Satan, the manipulation of man. The rules of the game, which necessarily include an element of creativity, imply that evolution be left to itself. And all these attempts of man to control natural interactions mean the total abolition of any creativity, the end of evolution and the end of development of human art, of human dignity, of human science, of all human values. So engineering whether social or genetic is certainly out.

M.S. Is living 120 years possible? It is desirable?

K.L. Yes, yes. I am 74 years old. Yes, even with my arthritis, my cane to help me walk, and other little miseries, if I am otherwise not changed, I should like to live 120 years. Just to see what happens. Using a yardstick of increased age, if 120 years should be normal, retirement would be at 100 instead of at 65, and you would retain your creative power, your youthful creative power until you were 70. And under these conditions it would be perfectly desirable to live very much longer. The human brain is able to collect a great deal more knowledge than we can make use of in our too-short life. And you know who said that: Bernard Shaw. In his play, *Return to Methusaleh*, he said just what I am saying now to this question.

M.S. Could euthanasia tomorrow, under social, political restraint, be part of a new morality, an ethic inherited somehow from the new knowledge we have in biology?

K.L. That is a question of personal responsibility. I know a number of cases where medical people killed people condemned to death by painful terminal illness. My brother killed his wife, who died of breast cancer at 28. Well, he gave her morphine, morphine, morphine without any consideration of prolonging her life, the only consideration being to spare her pain. And I know of many other cases, but I doubt whether you can institutionalize this. Human frailty is such that it brings out the fact that old odd Maria doesn't like life anymore, and her dear young nephew would inherit her millions, and he would do so much good with these millions, and she suffers anyhow, so let's perform euthanasia.

M.S. Since the second world war, there have been many problems which we cannot talk about without seeing the shadow of the Nazi past.

K.L. You don't need to tell this to me. There are some things you cannot talk about without being branded a Nazi. For some time to come, Hitlerian eugenics will go on paralyzing any spontaneous discussion on this topic. It makes life very difficult for everybody, we know that. We know that something has to be done about euthanasia, even if not through institutionalization, because not everybody is a saint; not everybody is a wise person.

M.S. In California there are attempts to work out a kind of a code of behavior and to palliate any institutionalization by kind of a biologic morality for physicians and professionals.

K.L. I am all in favor of that. That is what's necessary and, between you and me, I'm writing a book on that attempt to create a new morality.

M.S. That is such a problem, because people don't go to church, they are no longer restrained by their religion; or they are in moral conflict with their behavior: free love, contraception, abortion.

K.L. It is much worse than that. It is the psychotic attitude of the whole world today, what I call "scientism," and this is the belief that anything that cannot be defined in terms of the natural sciences and anything that cannot be verified or quantified has no real existence; this throws all values into the realm of illusions. Did you read Skinner? Freedom and dignity are illusions. Free will is an illusion. Even beauty is an illusion. Goodness is an illusion, and this is somehow an accepted creed today. In journalism you never find the adjectives good or bad, of decent men, of kindhearted men, that doesn't exist; it is all an illusion nowadays. That's old-fashioned. This applies of course to euthanasia. I should say euthanasia today is already much more widespread than you know. It is performed by the person who loves the dying man best, and he is the only one who has the right to decide. Of course modern medicine puts responsibility on the shoulders of the medical man, which ordinary human beings are not strong enough to bear. Say a man is in an iron lung, and you know if you switch off the power, he is dead. He can live for years; he can speak. Now another man is coming in who is only temporarily unable to breathe. You would be morally constrained to kill the hopelessly ill patient, who is your friend whom you have treated for years, whom you know personally, to put in a new young man who perhaps will recover. And that would morally be your duty as a medical man, and this is simply more than a person can bear.

M.S. In the United States and especially in the city of Seattle, on the west coast, there is a committee of concerned citizens that choses who should live and who should die. For instance, if there are five people who require certain technology, an artificial kidney or something like that, and there are only three available, this committee of concerned citizens has to make a choice. What kind of choice are they going to make? So we need a new code of morality; we need it because people are in search of rules, which they don't find anymore in the Church. It is what I call the necessity of a new morality of biology.

K.L. My new book is called *Evil*. My book on aggression was called *So-called Evil in the Natural History of Aggression*. And the subtitle will be "Attempt at the nature of morality," related to your subject.

M.S. There is a kind of gentle good euthanasia, modern man's conditioning by an array of psychotropic drugs. Do you think man should fear the loss of his free will, his freedom by these new psychotropic drugs?

K.L. I should say yes, I should say that I fear that. I fear that because I don't see how if one uses psychotropic drugs, one could guard against addiction. Can one take precautions within a democratic society against excesses of this nature? That would of course impair your freedom of choice, liberty. All these precautions are based on a perception of values, values that cannot be defined in precise scientific terms and that cannot be quantified.

The Ministry of Health can publish pamphlets. Physicians can be educated, informed of the consequences of their prescriptions. Excesses should be prevented. Amounts of the necessary psychotropes should be severely controlled, their sale should be judiciously restricted. Is that enough? Is it feasible? After all, people who take barbiturates are not acting illegally: a barbiturate is not marijuana nor is it cocaine. It's a very complex matter, and that is one of the prices we have to pay for the privilege of living in a democratic society: free access (or almost free access) to substances that may destroy us.

M.S. Can mental illnesses benefit meaningfully from manipulations of the psyche, from psychotropic medicines, or from electronic means?

K.L. I should say yes as far as drugs are concerned. Deep depression can be helped enormously. A real psychosis can be helped through electronic means; I have seen a schizophrenic really cured by electric shock, though this happens rarely. I saw such a patient, when I was a military physician. I brought him to a psychiatrist. This man was brought into my department with a developing schizophrenia. And I have had the good luck (or bad luck, as you might say) to watch the growth of a schizophrenia, and I can tell you it's the most horrible thing that can happen to a person. The nightmare quality of this man's life; he was really suffering. He had electric shock and snapped out of it, which happens occasionally. I could follow him only after he left the ward, but he had no relapse as long as I could still contact him, that was over one year.

M.S. What I meant to say here (because it's a kind of look at the future) is, what do you think of psychic manipulation done by new means, by computer or computing systems?

K.L. Is that what you mean by electronic means?

M.S. Yes. In Veterans' Hospital in Salt Lake City, Utah, I saw a man treated by a machine, in a way. A man comes into the ward and he is questioned by a machine which asks him a lot of questions including the more intimate questions about his sexual life. It's funny, but maybe only Americans can do this. The man I saw was answering the machine with great sincerity, even more so than if he were interrogated by a psychiatrist.

K.L. He was not ashamed?

M.S. Not at all, apparently. Of course, afterwards, he is seen by a psychiatrist, or a psychoanalyst.

K.L. I am rather ambivalent toward psychoanalysis. It has been very much abused, and it's too easy to earn money with psychoanalysis. Yet I believe there's a lot of truth in the basic Freudian ideas of the subconscious and of the healing effect of raising the subconscious to the level of consciousness. There is no doubt that this is true and that it helps. This is obviously one of the greatest discoveries in the treatment of mental illness since the beginning of time. At a time when the world was ruled by the idea of stimulus/reflex psychology, Freud saw that instinct is a spontaneous generation of something which requires effort to be kept under, to be suppressed. And if he had done nothing but that, he would have deserved three Nobel prizes. That's my opinion of Freudian discoveries. I take a very dim view of many psychoanalysts who make a lot of money, out of pseudo-psychotherapy; we would be much further in our knowledge of the human soul if Freudian discoveries were not so stupidly applied.

M.S. You're right. But every discovery gets commercially applied, including God.

K.L. Including God. I was just about to say that.

M.S. As for the sexuality of tomorrow, in a world of contraceptive vaccines and test-tube babies, will conception and sexuality be totally dissociated?

K.L. God forbid.

M.S. The process is on its way. We are only at the beginning: we have the test-tube baby and there are cases in which the bearer of a child is not the mother. All this with the best intentions. The fight against sterility . . .

K.L. Hell is paved with good intentions. The dissociation—for the sake of scientific efficiency—of the feeling of love and the natural reproductive process goes together with genetic control and manipulation, artificial insemination, etc. All this will cost a lot of money. These types of increasingly complex prostheses, which only the rich will be able to afford, will result in a horrible increase in propagation in the great mafiosi lobby industries. We must not forget that the sense of values, all that makes for good and beauty, is genetically fixed. And we do not know anything about genetic fixation; we do not know at all how it works. By controlling natural selection, controlling the natural choice of partners and so on by this dissociation, we might do a lot of harm. . . . It will be possible to dissociate love and sexuality, but if so, it would be highly dangerous to humanity.

M.S. You believe that the monogamous family is inscribed in our genetic code, rather than being the product of Judeo-Christian culture and tradition? But animals do not form couples, families, in the Judeo-Christian way.

K.L. Oh, yes, some species do. That is a very interesting question that Niko Tinbergen and I often discussed. If you observe the process of falling in love and if you read everything that has been written from Persian poets to Oscar Wilde or Thomas Mann, you would believe that man is a monogamous animal, because the oath of falling in love is something which points to a monogamous genetic code. On the other hand, if you just gather statistics in terms of who copulates with whom, you would come to the result that man is completely promiscuous, at any moment and with everybody. All these genetic programs are open ones, which are modifiable and which are realized in different forms under different ecological or environmental conditions. For instance, there are fish that, under certain rather unfavorable conditions, form monogamous bind pairs, and under more favorable conditions form harems. And I should think that this idea of pairing is instinctive in human beings, which at least temporarily makes for monogamous pairs. And maybe, also indubitably, the dissociation of being in love and wanting to copulate, these two things should not exist separately, and the linkage between them is certainly much stronger in females than in males. You will find very few men who are not ready to copulate without being in love, and you will find very few women who are ready to copulate without being in love. And this is exactly the same in geese. It is exactly the same in certain monogamous birds, while dissociation is much easier for a man. Now, if you look at the genetic programming of the child, you find quite unambiguously that the child has room for two parents. It must possess a father and a mother. And of course, you must not tell this to Women's Lib, because they will lynch you, but all these programming points put together would point to one male per family. A child with a mother and a father does not suffer if the father has other women about him; the Arab and many other polygamous societies work very well. And of course, women's answer is that the opposite is not true: there are very few polyandrous societies. But in Tibet, two brothers usually have one wife together and it usually works successfully as it diminishes jealousy. Then again, I have to say that the same happens in geese, where two brothers have one female together. With, however, a certain amount of homosexuality between the males. I have observed that only in geese and in Tibetans. For me, the dissociation of these things would be—though I am not a prophet—highly dangerous.

M.S. I am no prophet either, but what I predict is that we will most likely have a return to the most old-fashioned behavior in love, with romanticism.

K.L. I agree with you. There is a commune in Vienna (just as there are communes in America) which is composed of men and women, and where it is forbidden to copulate successively with one person. The rules of the community require copulation with every member of the community, and this community suffers from the fact that, despite their "religion," despite their intentions, a man and a woman stick together and leave the community to become monogamous. Old fashioned romance wins the day, as you rightly said.

M.S. It also started like that in some of the first kibbutzim in Israel. They decided to share everything, including the women. But it did not last.

K.L. They have evolved since the time of free love and Spartan separation of children from the family. You know, I have Israeli friends, and I am informed, to some degree, of the development of the kibbutzim. They get more and more intelligent. For instance, they have stopped trying to separate children from their parents, and the most modern kibbutzim insist on having the parents living within walking distance of the nursery, so that the children can visit their parents in a few minutes. One of the most interesting results of kibbutzim is that boys and girls are kept in groups up to the age of ten; they are not apart at all. They mix and bathe naked together and are completely immodest in that way, and it hardly ever happens that sexual behavior arises within one group. And if it does, the pair is ostracized and the pair judged against by the group itself, while love between groups is highly favored. A girl or a boy who is in love and who copulates with a partner from another group receives the approval of the whole assembly. That is so incredibly beautiful because here you have the programming against incest. You do not fall in love with your sister, and if you have been brought up with her, you cannot fall in love with her.

M.S. Those kibbutzim are then a kind of family? All are then sisters and brothers within a group?

K.L. Yes, and therefore, they react against incest, as they consider themselves as brothers and sisters, and consequently of no interest sexually among themselves.

M.S. Could one imagine a world where, thanks to prostheses and grafts, failing organs could be replaced as parts of an automobile are?

K.L. Yes, of course. We have banks of organs. But God knows what complications this might involve.

Suppose I have three kidneys and four applicants: I would have to decide whom I shall let die. That's no joke. The case has already occurred more than once.

M.S. In many countries, there is terrible resistance from public opinion, from churches, etc., against removal of organs, in the name of bodily integrity.

K.L. Emotionally, I can conceive that resistance. But intellectually, I am in favor of transplants. I think that immunologists will be in a position to prevent the action of the antibodies which causes rejection of transplants. When this technical obstacle is overcome, this will become an interesting subject. It is obvious that, if need be, I would give up my fortune in order to buy a kidney for my wife, if I could afford to do so from an organ bank. The problem remains of how to maintain those organs, through freezing or other methods. It is clear that if you have a sufficient stock of organs—even higher than the demand—then your problem is solved.

M.S. It could be solved even more easily. One of the major obstacles against organ removal is the law. Are you then in favor of organ removal and transplant?

K.L. I have absolutely no respect for the body. Suppose we were on an expedition to the South Pole or elsewhere, and we ran short of food. I just hope you wouldn't kill me to have a nice roast beef of Lorenz. But suppose I fall and kill myself, then go ahead, help yourself. I have not the least objection to your helping yourself to my liver or my kidneys. I wept on the accidental death of a tame goose that killed herself against an electric wire, but I ate her up: there was no reason to have a sinful waste of food.

M.S. Some of my friends, in California, carry a card indicating that if they are very sick, their lives should not be prolonged and if organs are needed, theirs are available to the Institute of Pathology.

K.L. I should say—avoiding any simplistic consideration—that it is too hard for a surgeon to decide who is to die and who should get the transplant. If the organ banks are in a position to meet organ requirements, in the interest of all who suffer, we must give up our emotional reactions. I am, therefore, all in favor of these organ banks.

M.S. Can one conceive, because of a population explosion or because of the exhaustion of natural resources, of a habitat for man in space or in the sea, in large communities?

K.L. Yes, one could. Or at least on satellites. But that is no solution. A child who has understood the principle of compound interest must understand that exponential growth in a limited, finite environment must lead to catastrophe. Therefore, when one makes this connection, one necessarily comes to the conclusion that we will end in disaster if natural resources are being exhausted, and commercial and population growth go on as they do. So finding new habitats would only postpone the problem for a couple of years and not solve it.

M.S. You mean it is technically possible but useless?

K.L. It is as useless as nuclear power. Atomic power is just one symptom of exponential growth. It does not solve the problem of energy, because you get energy from the gradient of the temperature and not from the temperature itself. And even if we used atomic fusion instead of fission, this would mean, in the long run, a heating up of the planet, which would make the deserts drier, and melt the ice cap of the poles, and drown Venice and New York. My sermon on atomic power is quite short. Any intelligent housewife knows that she cannot spend more money than the family earns. Otherwise, they are living on the capital. And the only legitimate income of earth is solar energy. There is no escape. Remember what I said about the child who has once understood how to calculate compound interest. We must face the fact that big business, the powers that be, have put themselves into a vicious circle: they must grow or go bankrupt. And the main task, which should be the sole concern of modern business, is the question of how to get an interest rate out of that, because exponential growth cannot do this. The expansion curve must be changed into a symptotic curve going parallel to the zero line. As to how to do that, I don't know. That is the task of the mathematical economists who must find a solution. But these are the fundamental problems of humanity today. The powers that be are completely immoral; that is what we must also face. Those in power are absolutely ignoring any but the one interest of earning money. The drive to make atomic bombs powers American industry as well as the Russian economy, and both are equally short-sighted. Nobody is interested in long-term plans for humanity. The interest of politicians and state employees never extends beyond the next election. These circumstances may be the death of humanity, and if I am sermonizing all the time, it is because I am an optimist. I believe that a solution will be found but that to do so is urgent and essential.

M.S. Can you imagine the day when immunology will replace chemotherapy and surgery?

K.L. I should say no, because we cannot play around with biological equilibriums; however, what immunology can do is to make organ transplants successful. Immunology is not the panacea of therapy, but sometimes we can look forward to universal answers, keys that open all doors.

M.S. Is computer-controlled health care a good thing or kind of a prelude to an even more police-like concept of tomorrow's society, one that would offer no possible escape?

K.L. We are on the way to that. Even now, we cannot escape, because a normal modern man has such a narrow field of what he can do himself. None of us could survive independently in the woods, without the help of experts. And the limited effect of expertise was made clear to me a few years ago, by my family doctor, Dr. Duyak. He came along with a lovely new electrocardiograph and I was enthusiastic about the beautiful little machine. Then he said, "Yes, I had a huge, old thing before; my old electrocardiograph, which I understood and which I could repair alone; but this new one is completely beyond me. If it fails, if it breaks down, I'm stumped because I cannot even attempt to repair it myself." And this applies to our daily tools. In my old car, I had a carburetor which was working and which I understood perfectly. My new car has a computer-controlled injector pump which cannot even be repaired in any workshop in Vienna: it has to be sent to the manufacturer in Stuttgart. So the things that we are given are fool-proof: we must not tamper with them; one must strictly follow the instructions on how to use them. And that makes you the slave of the expert. And technocratic expertise is a totalitarian regime that catches us all by the means of conditioning, by reinforcement, because it is so easy, so comfortable. Suppose you want to realize how much you are subject to coddling, to spoiling by commodities, by comforts; just visualize a house without central heating; no hot water, water to be heated on the stove; no electrical lighting, but just oil lamps. I still remember the rows of petrol lamps of my youth which were cleaned every day. The same thing applies to cars. Just imagine the old means of transportation and imagine a return to the old vehicles. This would be going from silk to straw. If it is very easy to go from straw to silk (men do this without even noticing, without even enjoying); going back from silk to straw represents the most horrible frustration. And that is an influence under which all in this world (except the very poor, unhappy third world people who still are exempt from this development) go on living. This comfort is weakening us. Its excesses act on us just like a drug, and we become addicts.

M.S. Computer-controlled health care management may also have great benefits. Don't you think that a democratic society has means to protect against the abuses you fear? After all, the Soviet leaders need no computers, nor do they need any sophisticated western devices to be a totalitarian society.

K.L. No, but computers and so on will sooner or later give us something very much like Soviet tyranny.

M.S. Do you actually think that European democracies or the United States will fail just because of the use of electronics?

K.L. I am afraid so.

M.S. Why shouldn't a democratic society control the computers instead of being controlled by them?

K.L. Because man almost always becomes the slave of his tools.

The more complex the tool, the stronger the slavery. I would not trust myself—nor democracy—to control the computer. I think the danger exists of the computer controlling democracy instead of serving it.

M.S. Will man be able to biologically control his own body, by means of microprocessors? Is it desirable?

K.L. It is feasible, but not desirable. Because all those things which allow for a control on one's own body should only be in the hands of those very clever persons who know how to use them and who know what they do. So far, biological mechanisms, which proved their value through a long evolution, are more to be relied upon than all these inventions. Knowledge is important, not the tool. All this would be all right if everybody were a doctor. Indeed, while it is easy to take one's blood pressure or make an electrocardiogram, the reading and understanding thereof are not as easy as they are believed to be. Thus, the use of those measurement and control devices implies a much wider knowledge than that of the average man of today. The alternative should be "Yes, if the 'controller' has the knowledge required for making a valid decision."

M.S. So many miracles are expected from the new biology that this discipline is already considered as being able to solve all our problems.

K.L. Why not? Given the condition that we are able to control exponential growth. And here we go again. Indeed, this brand-new technology may temporarily solve some problems, while, simultaneously, speeding up in the long run the process of catastrophe, by ever-increasing use of energy, because of population growth, etc. One of my friends, Hassan Osbekan, who is an economist in San Francisco, told me, "All threats to today's humanity arise from overpopulation, and the solution to this problem is education and nothing else." That is a very firm statement, and the more I think of it, the more I realize it is true. Indeed, you cannot require that a man of a developing country—who is starving and who is worried about his rice or other cereal crops—avoid using DDT, if he has DDT available to him. You cannot tell him that DDT, in the long run, kills everything. He would not understand it, and you cannot expect that he would understand. You cannot expect from a Pakistani or a starving Indian that they understand that our Western fat is not perfect happiness. Education of human populations implies, first of all, some equality in the standards of living. As long as there are people starving, you cannot ask them to stop using pesticides, nuclear energy, and God knows what. This means that *we* should spare as much as we can.

M.S. We will never reach the point where Western man—European or North American—will willingly accept living more frugally in order to give food to those who are starving.

K.L. Maybe we could give them more, without requiring any sacrifice from us. But anyway, we have to moderate ourselves. I once participated in a conference held in Tokyo, the topic of which was Man and the Sea. On this occasion, the Japanese minister of fisheries declared that it was wrong to think of economy and ecology in separate terms. Economy is nothing but a shortsighted ecology, and nothing can be economically right if it is not ecologically justified. To me, this is evident. At this conference, there was also a certain Krauen, an Austrian technician who was also referring to exponential growth. He said that the only hope of mordifying the greed of the rich is out of a catastrophe. He also added that, as long as we are privileged, we will not believe the reality of the danger threatening us. He also said that, as long as no major catastrophe touches a big city, one with a high density—in the U.S.A., in Germany or elsewhere—which would destroy a large part of industry and affect thousands of Westerners, there is very little chance that we will spontaneously limit ourselves. He shared your view that it is impossible to moderate the desires of some rich people in order to give food to others.

M.S. I think that this move toward more moderation can be achieved with an elite, with people who have values they can substitute for belongings such as a second car or a country-house.

K.L. That too is an educational problem. The manufacturers of consumer goods are being tricky when they bestow on these goods a symbolic value linked to social status. For instance, if Mrs. X has a color TV set, you are ashamed to have only a black-and-white TV set. The spoiling of man by comfort is obviously a danger. It is difficult to go back to a bike, after having enjoyed a car. No miracle is to be expected, as long as we have population growth and overexploitation of existing energy resources. The latter—whatever be their importance—have an end, a limit. Our planet has only a finite amount of energy resources, and the more the population grows, the more modest will have to be the needs of each individual. One must divide the world income by the number of people on this earth: the calculation is very simple. You might say that this is a utopian view of the future. In one way or another, if we want to survive, we will be forced to make plans, global decisions which seem utopian today.

M.S. How do you see the role of the physician and of medicine in tomorrow's society?

K.L. I stick to my view that the intuitive powers of our brain are more sophisticated than any computer. Our brain is capable of doing things which no computer could perform. If a computer comparable to our cerebral cortex were existing, with as many electronic elements as there are cells in our brain, this machine would have to be the size of a mountain. Therefore, I do believe that a good intuitive examination by a general practitioner—the clinical eye, as we call it in our jargon—cannot be replaced by a machine or any army of specialists. The general practitioner—who knows nothing—knows, however, a little bit about many things. He has an overall knowledge, and I think that this role of the old family doctor is absolutely necessary, even if my view may seem to be old-fashioned romanticism. He knows his patient; he knows the pathological history of the family. Confronted with identical symptoms, the doctor's reaction will be different according to what he knows about such or such patient. With a patient before him, whose history is known to him, he will see at once whether the patient looks paler, whether he is fatter than usual; and from there he can make a diagnosis that no laboratory analysis will ever be able to replace. Thus, I think that the role of the family doctor is purely and simply indispensable. As to medicine, I think it has already reached a stage beyond human control. We are already deciding who is going to live and who is going to die.

M.S. Then you have the patients' revolt. They want another kind of relationship with their doctor. They want to participate and they do not want to be just subjects who are given medicine without a word of explanation.

K.L. They want a friend, someone they know to be good.

M.S. Not only. They want a partner to deal with, someone open to dialog.

K.L. All this is included in my concept of the old family doctor who, of course, speaks to his patients. Therefore, when I go today to a hospital and the internist sends me to six various labs for examination, without even having looked at me, I am frightened. I personally prefer the old family doctor to the specialist, for whom the performance of his technical equipment—in the form of laboratory tests, electroencephalograms and electrocardiograms, etc.—seems to be more important than the patient. The patient is of less interest to him than the good working order of his technical equipment.

M.S. Can one conceive of a kind of preventive medicine that does not include coercion?

K.L. At the present stage of average human intelligence, I would say no. You cannot bring a man who does not feel sick to preventive medicine. You must have a certain amount of knowledge about cancer, for instance, you must be sufficiently scared of it, in order to have a regular medical examination to detect cancer. Consequently, I think that, at the moment, you cannot have preventive medicine without coercion, be it only for vaccinations, for instance.

CHRISTIAN DE DUVE
A Great Flemish Gentleman

To be a native of a little country like Belgium has advantages as well as disadvantages. All cultures and all currents of thought cross these Flemish plains in the heart of Europe—plains buffeted by winds from every direction and open wide to the Atlantic and the New World.

Christian de Duve, one of the most illustrious men of science in the West, is a citizen of the two worlds, old and new, since he teaches both at Rockefeller University in New York and at his alma mater, Louvain's Catholic University (in the French-speaking section, having been moved to the suburbs of Brussels as a result of linguistic quarrels). The latter is an eminently European and cosmopolitan university, and it is there that Christian de Duve set up his International Institute of Molecular Pathology.

His work, a matter of patience, attention to minute detail, and hypotheses long considered and verified "at the bench" of the laboratory, won him the Nobel Prize in 1974—awarded for his discovery of "lysosomes" and "peroxysomes," an important step toward understanding the inner workings of the living cell, of which these particles are a part. Christian de Duve's contribution in the fields of biophysics and biochemistry is exceptional, and the list of his publications—all scientific—is considerable. Plainly, de Duve is not fond of generalizations. He is strict. I can easily imagine him among the "heren" wearing ruffs in Rembrandt's "The Anatomy Lesson," playing the part of Master Nikolaas Tulp, the professor. But behind his forbidding appearance he is very generous, devoting himself to his students and to causes that seem to him just: funding for scientific research, and support for the dissidents in the Soviet Union and elsewhere. . . . This strict bourgeois from Flanders is a great gentleman.

M.S. Does the list of miracle drugs recently compiled by various scientific institutions strike you as a reasoned and reasonable forecast of what's to come, or as science fiction?

C.D. I have seen several such lists of drugs. Every five or ten years specialists are asked to predict the time when we will find a drug effective against cancer, certain allergies, etc.

All the lists I have seen have always proven to be too optimistic. Recently I looked over a document of that kind compiled in 1965. The predictions then made for the year 1980 are far from being realized.

We should be reserved in our predictions. But basically I am persuaded that sooner or later we will find solutions to most of our present-day medical problems.

To come back to your question—futurology or science fiction?—when it comes to curing or preventing disease, nothing strikes me as being really ideal, or utopian. But I have no idea of the time it will take to reach one goal or another. To attempt such prophecies would be presumptuous and naïve.

M.S. If we think back to the past, we may arrive at an indication of the time that has to elapse between an initial insight into a fundamental concept and its application, say, as a cure or preventive measure for a given disease, apart from the very special case of antibiotics.

C.D. In the past, most discoveries were pragmatic, the result of empirical studies, and therapeutic application often began very soon thereafter. It was about 1875 that Robert Koch managed to show conclusively that the *"bactéridies"* described by the Frenchman, Davaine, in the tissues of animals who had died from anthrax were in fact the agents of that disease. His was the first rigorous demonstration of the existence of an infectious agent. In only a few years, researchers would identify the microbes that cause a whole series of infectious diseases;

they would perfect vaccines and develop a large-scale policy of preventive medicine that is still the basis of present practice. It was in 1887 that Eykman associated beriberi with a deficiency of an essential nutritional substance—a vitamin. In a few years, vitamins A, B_1, B_2, PP, etc., were discovered. All this was done in quick succession.

M.S. Still and all, there have been cases where a rather long time elapsed between discovery and application.

C.D. There have been, and there still are, discoveries that perhaps come prematurely—whose importance or significance is not appreciated at the time, or of which advantage cannot be taken for technical reasons. Thus, more than forty years elapsed between the demonstration that diabetes was due to a deficiency of a hormone secreted by the islands of Langerhans in the pancreas and the first injection of insulin into a patient.

M.S. Take the case of interferon. We now know what it is, but we don't know its exact role. And probably quite a bit of time will elapse before we know both how to manipulate it and how to make good use of it.

C.D. Yes. But, once again, it is not our ignorance that for the moment is preventing us from using interferon but the fact that interferon is quite simply not available in sufficient quantities. Most drugs have been used—and many still are today—before anyone knew what they could do or how they would act. Take the case of aspirin: Millions of tons of it must have been consumed before we had any real knowledge of *how* it worked.

M.S. Let's consider the man of the twenty-first century. Do you see him as more congenial than he is now? Or will the population explosion, the increased density of the urban fabric, and the increased scarcity of natural resources make people more aggressive?

C.D. I am tempted not to answer that question. Of course, each one of us, as an individual, has certain opinions about his times, his contemporaries. But those opinions are seldom based on deep reflection. I may talk as an expert on biochemistry or cellular biology. But am I really capable of making pronouncements on problems such as the aggressiveness of my contemporaries, especially if I must compare it to that of other generations? The same thing goes for the population explosion and urban problems. . . . I think it's dangerous to put such questions to a man like me just because I'm a man of science who has made certain discoveries. That definitely does not make me more competent to judge today's society. I must admit that I'm a bit allergic to the disease that has afflicted many of my colleagues. I mean using the distinction accorded them for having done some precise scientific work to make pronouncements on the affairs of the world, as if they were augurs. To be a good scientist, one doesn't necessarily have to be wise, and even less to be well balanced. Many researchers I know are totally lacking in good sense and judgment.

But since you asked me the question, I will answer it, not so much giving you an opinion as telling you how I feel. I'm going to give you some examples. The generation of my children—that of the students of May 1968—seems in the eyes of my generation to be much more aggressive, because it challenges what we in our youth learned to respect: above all, authority. These young people have challenged the authority of their parents, of their professors at the university, of the government, of the police, etc. They are at war against all forms and manifestations of authority in the world, and sometimes in a rather violent manner. On the other hand, I believe they are much less aggressive toward one another than we were.

I get the impression that these young people want to eliminate from their world certain forms of competition and aggressiveness that were considered normal in our society. When I was attending school, it was very important to be the first in one's class. By awarding prizes and laurels, the entire structure helped to create a climate of competition that was regarded as healthful for the young. Today, young people are trying to eliminate that rage for competition—what the Americans so cruelly and so rightly call the "rat race." I believe, as these examples show, that aggressiveness remains but that the targets change from one generation to another.

As for the increased density of the urban fabric, the problem is unquestionably serious. As you know, I live part of the time in New York, a city in which a nightmarish evolution is taking place. We are witnessing the deterioration of certain residential districts that could end in a situation similar to that of London in the eighteenth century. In parts of town where no one dares go for a walk unless he is armed, there reigns a kind of internal law that is very close to the law of the jungle. It's rather frightening.

As for the scarcity of natural resources, this will, I feel sure, make people more aggressive. That seems obvious to me. Will human inventive genius make up for the exhaustion of our natural resources? I'm tempted to believe so. But that will not suffice. If we don't manage to check the population explosion by peaceful and rational means, it will finally check itself, but by the usual means of natural selection: wars, famines, epidemics. In some parts of the world, unfortunaturely, things are already at that point.

M.S. The successes of genetic engineering periodically provide topics for discussion in newspaper columns. Do you see, in those successes, the promise of a Golden Age or the specter of an apocalypse?

C.D. A few years ago, genetic engineering was a cause of fear, but in large part that was due to the naïveté of the scientists themselves. They had good intentions but were somewhat lacking in good sense. Their call for a "moratorium" on certain experiments, dictated by a praiseworthy concern for prudence, was taken up and blown all out of proportion by sensation-seeking journalists, politicians, ecologists, and all those who in today's world preach the distrust of science. The consequence was the establishment of draconian and absurd regulations imposing more precautions for manipulating the imaginary monsters that genetic engineering would give birth to than for dealing with dangerous bacteria and pathogenic viruses. But good sense won the day, and those regulations have been made much more flexible. Personally, I see no reason to suspect that genetic engineering will bring on anything like an apocalypse. But neither do I see in it the promise of a Golden Age.

It is certainly an important development, especially for basic research; but it could never be a universal panacea. My feeling is that genetic engineering will allow researchers and molecular biologists to study, in a detailed way, the mechanisms of regulation of genetic transcription and expression. Also, it will serve as a point of departure for a number of important achievements, such as the production of human hormones by bacteria, and the modification of plants capable of capturing nitrogen. . . . But from there to a Golden Age is quite a distance.

M.S. Will people be able to live 120 years? Is it even desirable?

C.D. We should be able, first, to determine the degree of longevity inscribed in our genetic program. Most biologists think that death is programmed just like embryonic development or puberty. Thus, the normal longevity curve of a human population should be a horizontal line followed by an S. The sigmoid that would complete the curve would describe an average longevity (the age beyond which 50 percent of the population survives) on the order of from 90 to 100 years, perhaps more. Such a development is difficult to predict, and no doubt the key age would vary from one population to another.

Current longevity curves do not take that ideal form, but they are coming closer and closer to it. There are obvious reasons for this. Infant mortality has been considerably reduced, and we have eliminated a large number of infectious diseases (smallpox, tuberculosis, polio) and epidemics. Nutritional deficiencies are being corrected, along with hormonal disorders. Such pathologies have been conquered in our part of the world, where people are now dying of arteriosclerosis, cancer, and other so-called diseases of civilization. But when cancer and other degenerative diseases are cured, we shall increase life expectancy even further.

Finally, in a world characterized by a perfected medical science and the absence of accidents, everyone would live out their time and have a natural death—perhaps a very gentle death.

Old people who die at an advanced age usually have a serene end. They experience a sense of having reached the normal term of their existence.

Another approach, which is not to be ruled out, might be initiated by a change in the genetic program itself or in the conditions of its development. Perhaps one day we will manage to manipulate that program or to act on the environment in such a way that the genetic/environmental reaction would result in increased longevity. Experiments on animals give us reason to hope for this. The longevity of rodents can be increased 30 percent by undereating, as McCay's classic experiments have demonstrated.

To live 120 years would certainly be desirable to the extent that everyone could live in good health to an advanced age. There would be, however, no question of artificially maintaining the life of worn-out organisms and softened brains. I believe that medicine would be rendering a very dangerous service to society if it achieved this kind of "triumph."

M.S. Can euthanasia be integrated into tomorrow's new morality, under sociopolitical constraints?

C.D. Before saying whether one is for or against euthanasia, one must define it. We recently had the famous case of Karen Quinlan in the United States. Then there was the case of the monk who was hooked up to a series of machines that allegedly kept him alive—a purely vegetative life. The doctors finally decided to disconnect the system. Can we say that the termination of these artificially prolonged agonies is euthanasia? Or isn't it?

M.S. Most doctors don't want to get into these questions. They say one must proceed "case by case." True, every doctor has more or less practiced euthanasia, especially in hospitals. But do we need a code? Do we need regulations, with all the interference (sometimes undesirable) that a public authority would bring with it? Or should we continue to act empirically, taking into account our own conscience, the opinion of the collegium of doctors, the circumstances of time and place, the facts of the case, etc.?

C.D. I believe we're going to move from the present empiricism toward a certain codification, as has happened in all comparable cases. Society evolves, adapts itself, and takes on new laws and sometimes even new moral or ethical attitudes in the face of dramatic problems. It's plain to see that society changes in this way. All I need cite by way of proof is the recent evolution of mentalities and legislation in such matters as contraception, abortion—and soon, no doubt, euthanasia.

M.S. Still, the debate over euthanasia is much greater in scope than its strictly intellectual or moral aspects. I mean this in the sense that the conflict between the partisans of high-tec medicine (whose relentlessness leads to the debate about euthanasia) and those of medicine for the masses (the best possible care for the greatest number) becomes an economic issue in the field of health care. I don't believe that this is a strictly theoretical debate. I believe that it is a debate very much at the heart of the preoccupations of our time—a time of crisis when health-care expenses are becoming a very heavy burden for our national budgets.

C.D. I am aware of the seriousness and importance of the choices to be made in the matter of health-care policy. And I am much concerned by the problems of medical research. Which orientations should be favored? Which strategy should be adopted? Here, I think, we have taken the wrong road. We are developing more and more technologies that are at once costly and not very effective. They are what Lewis Thomas, who is very eloquent on the subject, calls "halfway technologies." Our hospitals are bankrupting the country with equipment that is more and more sophisticated but whose use and maintenance are becoming problematical. Budgets are overblown, and we are heading for bankruptcy. Society cannot continue to pay for the kind of care we are now dispensing in hospitals. Why? Because it is a kind of care that is too expensive for what it contributes. Many people have said this in print, but the message is not getting across: We are continuing to encourage the development of these halfway technologies. Thus it is that artificial kidneys, heart transplants, and all kinds of other spectacular techniques are bringing us inevitably to an impasse. On the other hand, we have the classic example of a *simple* technology that responds well to the problem it was created to solve and spares society considerable expense: the polio vaccine. If, in the matter of polio, we had followed a strategy comparable to that applied to the introduction and use of the artificial kidney, we would now have increasingly improved iron lungs plugged into computers. We would have developed electronic prosthetic devices to help people use otherwise lifeless muscles. Moreover, we would have wrecked the health-care budget—whereas vaccination against polio costs almost nothing. I am perfectly aware that much is at stake in the choice of a health-care policy. But, in the end, everything depends on the orientation we want to give to research.

M.S. Do people have reason to fear that their free will, their personal freedom, will be threatened by the new psychotropic drugs? Can we, within the framework of a democratic society, take measures against excesses in drug therapy? And, if so, how?

C.D. It is a truism to say that the abuse of psychotropic drugs can be dangerous. We have ample proof of this on the individual level alone. One hardly need mention the methods used by the secret police of various countries. But I really don't believe that we will have, in the immediate future, a political or other governing body capable of using psychotropic drugs on a scale that could affect an entire society—condition it as a whole. There is nothing new about psychotropic drugs: Opium, alcohol, and tobacco have existed for centuries. Then came the tranquilizers, the stimulants—the pills that "help one to live," as they say. Of course, there are abuses. Unfortunately, it is a disease of our century, a fact of our times.

M.S. Then, are we stuck with a laissez-faire attitude, in spite of its leading to so many individual tragedies? Marilyn Monroe is a symbol for all those who spend their lives caught between stimulants and barbiturates until death comes. Must we be so lax that we allow people to gulp down a whole assortment of these "pharmacological sweets"?

C.D. I completely agree with you. It's terrifying. But is prohibition the remedy? The American example in the past thirty years has not been conclusive in that respect. And one may well ask oneself today if heroin would be so rampant were it not the source of illicit gain. For that matter, psychotropic drugs *can* have a positive aspect. I don't have to remind you of their role in the treatment of epilepsy, certain mental illnesses, and other conditions. It was these drugs, for example, that made possible the discovery of endorphins and other endogenous psychotropic substances manufactured by the brain. An important advance in knowledge. On the other hand, altering one's states of consciousness can be managed easily without psychotropic drugs. Television, radio, and education suffice. I recently visited a totalitarian country. There, they have no need of drugs to condition the masses, and especially the children. All they need are teachers and leaders of youth movements who repeat the same slogans to children, beginning when they are five. The conditioning is psychological, and it is very well done, unfortunately.

M.S. What kind of sexuality will we have tomorrow, in a world where, from the contraceptive vaccine to the test-tube baby, contraception and sexuality will be totally dissociated?

C.D. In the foreseeable future, conception will in my opinion remain totally linked to sexuality, I do not believe that the test-tube baby will be with us tomorrow. Until we have proof to the contrary, babies will be conceived only in the course of the sexual act; hence conception will remain linked to sexuality.

You are alluding, I suppose, to the converse—the fact that *sexuality* is becoming more and more dissociated from *conception*. I'm not a moralist, but this strikes me as an excellent thing. One unique trait of man—what distinguishes him from almost all other animal species—is the permanence of his sexual availability. In man, sexuality is not genetically tied to the simple needs of reproduction. As for the laxity we find in society today, I see it as an episodic phenomenon. I believe there have been periods when people were puritanical, and others when they were more tolerant. Libertinage has always existed, as has the spirit of the Victorian era; and they have succeeded each other with variable rhythms. The pendulum swings throughout history. Generally, an excess of severity produces reactions of permissiveness, up to the point where the resulting excess of permissiveness causes a reaction toward strictness. It would not surprise me if our grandchildren or our great-grandchildren re-created a society that would again become very puritanical in reaction to the excesses of their parents or grandparents. For that matter, I get the impression that this phenomenon is already taking place. We are witnessing the beginning of a kind of return to romanticism, as evidenced by a revival of twenties' fashions, a bucolic conception of ecology, etc.

M.S. Do you think it possible that one day immunology will succeed where chemotherapy and surgery have failed?—except in cases of trauma?

C.D. I am tempted to believe—perhaps because it complements the direction of my own research—that it will instead be chemotherapy that will make up for the shortcomings of immunology. According to some present-day thinking, any development of a cancer implies a deficiency in the immunological mechanism normally responsible for the recognition and elimination of the cancerous cells. The latter are then free to multiply. From this perspective, any intervention that would reinforce the vigilance or destructive power of our lymphocytes over the cancerous cells would be very useful. But up to the present time, attempts at treating cancer with immunotherapy have yielded rather disappointing results.

M.S. Is it vain to hope we'll be able to extend the frontiers of serotherapy and vaccinotherapy?

C.D. Vaccination is a possibility. But we still have far to go. Actually, it is no more a question of vaccinating against *the* cancer than it is of vaccinating against *the* infection. Just as there are many kinds of infection, so there are many types of cancer. A specific vaccine should correspond to each type of cancer. It may even be the case that each patient should have the right to his individual autovaccine, prepared to order against his own tumor, just as certain patients were once immunized with pus from their own infections. Perhaps that can be done some day. One of my former students, Thierry Boon, who now heads the Brussels section of the Ludwig Institute of Cancer Research, has obtained remarkable results in that area. By inducing a mutagenesis of cancerous cells, he raises their immunizing power to the point where the animals inoculated with them not only do not get cancer but even become resistant to inoculation with nontransformed cells that, in the control animals, caused cancer in every instance. The long-range hope aroused by this exciting research is to arrive one day at similar results in the treatment of cancer in humans. In the meantime, I believe we must continue vigorously to develop chemotherapy, putting to maximum use the new discoveries in cellular and molecular biology. That's what another of my former students is doing on another floor of the institute. This man, André Trouet, has developed a new approach to chemotherapy, taking advantage of our knowledge of the lysosomes, those little cellular "stomachs" discovered in my laboratory twenty-five years ago.

Classical chemotherapy tries to create molecules capable of discriminating between a cancerous cell and a healthy one. In my opinion, such discrimination will always be difficult because there are too many similarities between, for example, a leukemic cell and a normal medullary stem-cell for one to be able to exterminate the former without also destroying the latter. Rather than continuing to search more or less empirically for the miracle molecule, we take known cytotoxic agents and provide them with a probing head, which will serve to make them enter, selectively or preferentially, the target cells. This approach is interesting because it is based on clear and precise scientific data and can be followed in a perfectly rational manner.

On the one hand, we know that the different types of cells, including cancerous cells, carry on their surfaces specific receptors that capture, by pinocytosis, certain well-determined substances. Also, we have an entire technical arsenal making it possible to identify those receptors and thus to recognize which of the molecules they select might serve as specific vectors for the cytotoxic agents.

On the other hand, we know what happens to the molecules captured by the pinocytic receptors: They are directed toward the lysosomes. So that one can, on the basis of the properties of the lysosomial enzymes, choose—for purposes of attaching the toxic agent to the vector—a chemical bond that will be split up in the lysosomes, liberating the drug in an active form. It is Ehrlich's old dream of a "magic bullet." But it is no longer a dream. Already, hundreds of patients have been successfully treated with the help of the first vector-drug complexes prepared in our laboratory. And that is only a beginning. Right now Trouet has in hand new complexes that could revolutionize chemotherapy. And he isn't the only one. Today, many laboratories are interested in this new approach.

M.S. In reality, chemotherapy is a form of immunology. It's a marriage, so to speak. In orienting a chemotherapeutic molecule toward a precise target, by means of a vector, the distinction between what depends on immunology and what depends on chemotherapy is blurred. This is considerable progress compared to the blind chemotherapy that we now use.

C.D. The immune system's functioning and the form of chemotherapy we are trying to perfect have something in common: Both depend on the presence of receptors on the surface of the cells. Thus, I believe that an avenue of basic research that should be developed and supported much more than it is today is the one aimed at better identifying the antigens on the surface of cells.

M.S. Does it seem to you that the management of health care by computer could be the prelude to an inevitably more policed society?

C.D. Throughout history, people have refused technical progress for fear of the possible consequences. But the fears are seldom justified. Yet, for all that, I am not optimistic. I'm convinced that the future is going to bring man face to face with very grave tests—tests that in one way or another will be linked with the abuse of certain kinds of scientific and technical progress. But again, there's nothing new in that. . . . It's always happened that way. Man has always used—for better or worse, for his benefit or his damage—the technology available to him. And I don't see any reason to believe that this won't happen in the future. Because, for it *not* to happen, humanity would have to acquire, very rapidly, a heavy dose of wisdom. And today's world would not seem to warrant hope of this happening. Yet, things are getting dangerous. I believe that in the foreseeable future we shall continue to be not the agents of natural selection but its toys. Wisdom will come, but at the end of how much time—how many centuries or millennia? It is impossible to predict, but it will come because it is necessary. If we don't wise up, we will cease to exist. Things will evolve in one direction or the other. If our intelligence is to be, in the end, the cause of our destruction, then we are the victims of a lethal mutation. If, to the contrary, our intelligence one day brings us wisdom—more than it is bringing us now—we may finally find a solution to our problems.

In any case, I don't believe that the solution lies in banning or strangling certain forms of basic research on the grounds that the ensuing discoveries could be used to the detriment of humanity. It is totally impossible for us to foresee what discoveries will be made, and hence what their consequences will be. Discovery is, by definition, unforeseeable.

Of course, once certain discoveries have been made, one can and should try to channel their application so as to avoid abuses. And efforts *are* being made in that direction, even if they are being made in a very primitive and crude way. Take war, for example. People are trying to limit the use of certain weapons. A while back it was poison gas; then it was biological, or bacteriological, warfare. Now it's nuclear warfare. . . . And look at the environment. Today people are concerned about it. That wasn't the case thirty years ago.

M.S. I have my doubts as to the nuclear question, based on its irreversible side effects. We have created certain *faits accomplis,* which carry with them problems—radioactive waste, for example—for which we have no scientific solution. It's in this context that our faith in science is most prone to be shaken.

C.D. That's true. But here again, I don't think the solution lies in "less science." Rather, I think it lies in "more science." The breeder reactor produces less waste than the classical reactor. And, when fusion has been mastered, the problem of wastes will no longer exist.

I must admit that so far I am not antinuclear. I am for the protection of the environment—for an extremely vigilant policy. But I'm struck by the fact that the technology people are so dead set against today has virtually not caused a single death so far. Of course, I'm not talking about Hiroshima. . . . The most serious accident at a nuclear power station was the one at Three Mile Island, and not a single person was killed. Why this violent refusal of a technology that we can't do without?

Needless to say, it is better to take too many precautions than too few. But in that case we should do away with the automobile, because in one day it causes more deaths than the peaceful use of nuclear energy has caused since its introduction. And we should also do away with tobacco: It kills millions of people every year.

We have to be serious. I'm not saying that we should not regulate the use of nuclear energy. But there is something absurd and irritating in the way people blame nuclear energy for the damage that it *could* do, while tolerating so many other aspects of our civilization that *are* damaging—not only potentially but actually. It's all part of the flare-up of irrationality that is so widespread in the world today.

M.S. Is there not at the same time, as this flare-up of irrationality might lead one to suppose, a renewal of religious fervor?

C.D. I don't agree with you there, because religion is a system that is perfectly rational once one grants its premises. The irrationality that is in favor today has nothing to do with faith or the belief in certain verities. . . . It is an irrationality linked to the total absence of intellectual discipline. . . .

Unfortunately, children are being less and less educated to conform to discipline, be it physical or intellectual. People want to avoid effort and eliminate rules, which some educators feel are traumatizing for children. I hope there will soon be a reaction against this attitude. Because a person can't learn to think correctly without making a concerted effort, without rules that are imposed from outside until one has learned to impose them on oneself. For me, that's where irrationality comes from: from a feebleness in the critical use of reason; from a lack of rigor in thought.

M.S. People look to the "new biology" for so many miracles that some are already counting on its providing answers not only to our therapeutic problems but to nutritional, energy, industrial, and other needs in tomorrow's world. Do you believe this possible?

C.D. People are often tempted to exaggerate. Plainly, it's always dangerous to promise miracles—solutions to any and all problems. The answer to your question is twofold: *no* for the miracle, but very definitely *yes* for the prediction that biological research is going to enable us to find solutions to a great many problems. I'm convinced of that. We are for the first time entering an age in which we are beginning to understand the phenomena of life—not only understand them but analyze them by more and more powerful means, and also control them.

Up to the present, we have manipulated living matter in a way that is essentially empirical and pragmatic, and we have succeeded surprisingly well. Agriculture, stock breeding, medicine, the fermentation industries, along with the agronutritional and pharmaceutical industries—practically all progress that man has made in exploiting or manipulating living matter has been the result of empiricism.

Now, these results have been achieved despite the fact that our knowledge of molecular and cellular function has been rudimentary. Thirty years ago the living cell was unexplored, a kind of *terra incognita,* or black box. Today we have opened the box, taken it apart, and understood a number of its inner workings. We know the structure of DNA; we know how to make a copy of it and introduce it into a bacterium, obliging the latter to make the product coded by the foreign molecule. Witnessing this revolution has been a prodigious experience. We have acquired so much information that progress in the future is bound to be spectacular.

M.S. Aren't you perhaps a bit too optimistic? . . . How much time elapsed between Stanley Cohen's initial research at Berkeley and the production—which is still crude—of human insulin by means of genetic engineering?

C.D. The molecular tool that serves as the basis for genetic engineering and has made possible its development consists of "restrictive" enzymes that cut DNA at very specific points. The discovery of the first restrictive enzymes goes back about ten years. When you think of the progress that has been made since, you are struck by the rapidity with which knowledge is modified and renewed. I believe very much in the future of biotechnologies. It is that conviction that prompted me to found the International Institute of Cellular and Molecular Pathology at the University of Louvain.

M.S. That was an act of faith in the progress of basic biology in the service of medicine.

C.D. Yes, if you will; but it was a reasoned act of faith. The credo that inspires us is that the "new biology" will help us solve very concrete medical problems. For me, this was a relatively recent conversion—dating from scarcely ten years ago. All my life I've done what is called basic research,—what one does out of curiosity, just to understand. I have always defended—and I continue jealously to defend—the researcher's right to search freely. And I have insisted that it is society's duty to provide him or her with the means to do that, because basic research is an essential step for humanity.

But the counterpart, with regard to society, is the researcher's duty to develop, or to help develop, beneficial applications of knowledge and of new discoveries when he glimpses them. In that connection, I condemn the arrogant snobbery of the "pure" scientist entrenched in his ivory tower and making it a point of honor only to satisfy his curiosity and not to do anything that might be regarded as "applied" or useful research. . . .

Biology is at a turning point. Thanks to recent developments in biochemistry, cellular biology, molecular biology, genetics, immunology, etc., the revolution has taken place. The time has now come for the application of its gains. All of these new kinds of knowledge must necessarily lead to major breakthroughs. But for that to happen, the researchers themselves in these various fields must agree to put their competence into the service of medicine. It is this conviction that prompted me, along with a few friends who thought as I did, to found this institute. It consists of a central trunk of multidisciplinary basic research on which are grafted units of applied research: work on chemotherapy, cancer, leprosy, malaria, sleeping sickness, rheumatism, hereditary diseases, auto-immune diseases, the control of fertility. . . . All of this is done at our little institute. And I did the same thing at Rockefeller University in New York. I brought together, under the same roof with basic researchers, a series of groups working on arteriosclerosis, immunization, aging, and other subjects.

We are already quite encouraged by our work and its harvest of positive results. But we have also run into some disappointments, the main one being the lack of understanding—if not the hostility—we have encountered from certain traditional medical and pharmaceutical groups. I suppose their reaction should have been expected. But I must admit that it surprised me.

I've learned that in science, as in other areas of human endeavor, there are clans, territories, different ways of thinking and of approaching a problem. Medical research has its own strategy, consecrated by innumerable successes and adapted to the type of education received by those who practice it. The same goes for pharmaceutical research. And, of course, it is also true of basic research, which itself is divided into distinct sectors separated by compartments that are almost watertight.

But that can't last. The future belongs to interdisciplinary hybridization—to multidisciplinary collaboration. Everything we do is based on that conviction. What must be avoided at all costs is out-and-out specialization. So it is that pretty much everywhere in the world people have founded institutes of specialized research—in cancer or tropical medicine, for example. In my opinion, this is a mistake, because such institutes seldom have at their disposal a critical pool sufficient for the basic disciplines. Our conception was different. We were concerned above all with assembling that critical pool—biochemists, immunologists, cellular biologists, and molecular biologists. That trunk of highly varied orientations invigorates all branches of research—cancer and tropical diseases as well as many others. We start with the principle that what is basic serves everything. I mean to say that there is no fundamentally exclusive base for any single category of inquiry.

M.S. François Jacob said one day that there is no cancer research. . . . I interpret his sally to mean that cancer involves the whole of cellular pathology, and that research in oncology is part of everything basic.

C.D. That's right. And I would add that the same thing goes for all other fields of medical and therapeutic research. I come back to the advantage of nonspecialization. Cancer, malaria, diabetes, rheumatism, congenital anomalies—all these distinct categories of pathology overlap and are mixed up with one another at the levels of the cell and of the molecule. In creating bridges between specific concerns and the fundaments, we are also creating many footbridges among them. Our experience has shown us that after scarcely five years, the unexpected contacts that result from this network can be amazingly stimulating. Thus, the idea of doing research—at the same time and in the same institute, virtually in the same laboratory—on chemotherapeutic agents against cancer and against parasitic diseases has proven to be extremely fruitful.

I must emphasize that the bridges and footbridges I mentioned are authentic organic links. Actually, it is the researchers themselves who establish them. We do not have two categories of researchers labeled "basic" and "applied"—only one. Each person is integrated into the common trunk of basic researchers and also takes part in one or more programs oriented toward a specific area of pathology or therapeutics. Thus, each of our researchers embodies in a way the symbol of our institute, a hybrid of the Esculapius's snake and Watson and Crick's double helix.

RENÉ DUBOS
A Bet on Man

HE WAS THE DEAN of prominent Frenchmen in the United States, since his discovery of the New World dated from 1922. He came to medicine via agronomy; and he was one of the pioneers of antibiotherapy, having discovered in 1939 two antibiotics, gramicidin and thyrocidine, which unfortunately could only be used externally. He was one of the first both to address the issue of our ecological interdependence and to demystify "naïve ecology." We are destroying not only our natural environment but also the subtle balance that resulted from the planet's earlier "remodeling" by our ancestors—a planet that we have for some time now been ravaging indiscriminately.

A half century of life in the United States did not affect his very French way of being at one and the same time ingenuous, tender, and rough. He was a bit the sly farmer when he took me on a tour of his department at Rockefeller University as if it were his back forty. Although retired, he was still the guiding force behind a team of researchers in that department. Here we see the pragmatic genius of the

Americas, which allowed a man as brilliant and prolific as Dubos to continue his work in a familiar setting and with appropriate means.

René Dubos was famous in the States. A foundation has been named after him; his opinion was solicited by the media; and the princes who govern the American empire often called upon him.

He was a wise and lucid optimist whose voice was more valuable than ever in our disabled world.

M.S. What is your definition of futurology as it applies to the medical profession?

R.D. I always ask myself the question: What has medicine done in a practical way to make life better for people? My reply is a general one: People are always talking about curing diseases, but in fact there are very few diseases of which one is cured, even today. One can cure certain infectious diseases but not diabetes. On the other hand, one can help people to live with their diabetes. And I use this example to illustrate my conviction that most of medicine has little relation to cures in the proper sense. The *restitutio ad integrum* is virtually nonexistent except for certain infectious diseases and some rare surgical operations. The great majority of therapeutic procedures help people to live with their diseases—to co-exist peacefully with them. I realize that is not the point of view of out-and-out optimists. I have a very good friend, an oncologist of international reputation, with whom I'm in constant intellectual conflict. He claims that medicine becomes important only when we understand the exact mechanism of the disease on the cellular or even the subcellular level, because only then does it become possible to intervene and eliminate the affliction. To give you an example, he was interviewed recently, and was asked: "Do you think we will ever find a cure for cancer?" His reply—which I knew in advance from having thoroughly discussed the subject with him—was as follows: "Yes, the cure will come—not from medicine but from purely biological, basic research."

In a recently published book, *Cancer, Science, and Society*, the author points out that the incidence of different forms of cancer varies enormously from one country to another. For example, stomach cancers have disappeared from our midst, while lung cancers have increased; and there are a great many examples of that kind. This proves that, whatever the mechanism, cancers are to a large extent associated with

René Dubos died in 1982.

the presence of a noxious agent in our environment: in the case of lung cancer, cigarettes; in the case of stomach cancer, certain irritating substances. In these two examples, the problem is posed differently. For the optimist I mentioned just now, the solution lies in understanding the inner mechanism of the transformation of the cell. For the epidemiologist, it lies in studying the life experience of people to learn why certain forms of cancer have disappeared while others have begun occurring more frequently, and to rearrange that way of living in light of those studies. I am inclined toward the latter viewpoint. It's not that I'm against molecular science. In fact, my old department at Rockefeller University is entirely devoted to that field. But the more I look at what is being done in medicine, the more I realize that in most cases intelligent medical intervention consists of helping people live with their diseases. And from that viewpoint we have had extraordinary success.

As you see me now, I believe I am in good health for my age. But when I was very young I had acute rheumatoid arthritis that left me with a very pronounced cardiac lesion, so that naturally I had to arrange my life to live with it. I am seventy-nine and very active physically. A year ago, as a result of excessive physical strain, I went into an auricular fibrillation, and I'm still plagued with it now. The means for enabling an individual to function normally until the age of ninety, even with a chronic disease, strike me as increasingly available. This empirical approach to medical research is totally opposed to that of the absolutists—those who believe that radical solutions will come from basic research alone.

M.S. When it comes right down to it, you dismiss the "utopia" of total cures. For you, futurology is reasonable empiricism. . . .

R.D. Yes, and it's very curious how I came to this way of thinking, because in the beginning I was quite a proponent of basic research. I have spent the greater part of my life in the laboratory looking for solutions that would bring about the utopia of total cures. And, as it happened, in the field in which I did most of my clinical research we almost reached this utopia, with the curing of lobar pneumonia by antibiotics.

But once outside the laboratory, I realized that many people around me were afflicted with a disease of which we knew neither the origin nor the evolution, and for which no treatment existed. From that time on, I changed my attitude toward medicine.

M.S. Meaning that you don't believe in miracle drugs?

R.D. I see the chemical molecule that prevents the contraction of pernicious anemia, and the digitoxin that keeps my heart in working shape, as miracle drugs of a kind. I conceive of them as capable not only of curing but even more of permitting an approximately normal existence.

M.S. What do you think of psychotropic drugs, which not only help people live but can improve their memory, affect their libido, or make them euphoric?

R.D. I have read quite a few things on the subject, and I'm a little more skeptical than most people on this matter—not about the effectiveness of the drugs but about the real possibilities of manipulating the individual. Human beings have a capacity for resistance that renders even the most powerful psychotropic drug useless in undermining their free will. I've seen for myself the extent to which men and women of strong character control their behavior and attain their ends against all kinds of obstacles. I express, here, my exaggerated confidence in man's power to control his life himself.

As a matter of fact, I'm going to address this point in a series of films in which I set forth my life. The series will be divided into three parts. The first part will be called "Development." I use as its illustration my own case, but apply it to somebody else. I was born and raised in a certain village in France where I lived until the age of thirteen. It is plain that biologically and mentally I was molded by that environment. If I had been brought up in Japan or Greece, I would be different. Thus, there is the purely biological aspect of life—determinism.

The second part of the program will be what I have called "Choices"—the choices one makes in life.

Why did I decide at the age of twenty-two to come to the United States, considering that I was not having any particular difficulty in France? All kinds of ill-defined curiosity, things I had read that perhaps gave me certain illusions about the United States, and a certain spirit of adventure brought me here. The second program has to do with free will—the fact that one can choose.

The third part concerns our extraordinary power to change our lives if we really want to, just as countries and civilizations have the power to transform themselves. One sets off on a certain path and at a certain moment says, "This is not the direction I want to go in." So one goes back and sets off in another direction. Such social evolution is completely different from Darwinian, biological evolution. . . . Darwinian evolution is irreversible. But the social evolution I refer to can, to the contrary, be completely reversible.

What I have to say may seem very optimistic, but it is based on

A Bet on Man

many examples of individuals who, despite all obstacles, have accomplished remarkable things in the course of their lives. Think of the Russian dissidents or those people who survived Nazi extermination camps. Those are extreme cases. But there are also those people who simply face up, with a tranquil courage, to the burdens of everyday life. Wasn't it Péguy who talked of fathers of families as the true adventurers of the modern world? . . . Many people have found it possible to resist the temptation to become dependent on drugs.

M.S. What do you think the man of the twenty-first century will be like?

R.D. I think that the greatest transformation of our time resides in Western societies' having begun to acquire the power to imagine in advance the distant consequences of their actions and their policies, material as well as social. . . . The simplest illustration, and the most fashionable one, is the energy crisis. Actually, a country like the United States has much more energy than it could use before the end of the twenty-first century. And yet people everywhere are beginning to act as if we are really struggling with serious depletions. People imagine remote consequences in advance; they make up scenarios of the future.

Right now, the discussion no longer centers on whether our energy supply will be based on the atom or coal. My point of departure is a hypothesis shared by many scientists: We can perfect *many* methods of energy production for the next century. Moreover, we will succeed in controlling radioactive emissions. But there is one proviso with which everyone agrees: It will be necessary to concentrate the reactors in certain zones so as to be able to establish secure military and police protection of the atomic stockpiles. The aim will be to prevent accidents and ward off terrorist attempts. This conception supposes the acceptance of a very centralized state, perhaps even of a militarily organized society. At the other pole, the choice of solar energy would allow us to envisage a completely decentralized society.

These questions, which are the subject of heated discussion since they lead to opposite conceptions of social organization, are in my opinion the most important our societies must face. Are we going to choose the effectiveness of maximum centralization? Or do we favor respect of individual liberties?

I believe too that, despite appearances, people are less aggressive than they were in the past century.

M.S. Could a city like New York continue to exist in a climate of energy shortage?

R.D. Why not, if its energy consumption becomes more restrained, if it tries to reorganize itself? It would be difficult, I know. It so happens that a short time ago we organized a symposium, "Living Better with Less Energy." There, I said, in effect: "I believe that in reducing, even a little, the per-capita consumption of energy, we can improve many aspects of our life." Plainly, what I had to say applies to our Western countries and not to China or Africa. I am very struck by the reversal of the position taken earlier by officials of the White House's research service on energy for tomorrow. In an article published in *Science* and entitled, "How Should We Organize the Production of Nuclear Energy?" these people acknowledge that they had to give up their original idea of establishing many nuclear power stations throughout the country and opt for the concentration of nuclear energy production in a limited number of centers—nuclear megalopolises, as it were. The authors express more than a mere wish, describing in detail what to their eyes is a veritable program of action, although it is totally rejected by students and university people.

M.S. Is it possible to live for 120 years? Is it desirable?

R.D. As far as I'm concerned, I'm convinced that a person can live to the age of 120, and that we are on the point of acquiring knowledge that will make it possible. Personally, I would like to live that long. But if everyone is to live to the age of 120, a total transformation of our social structures will be necessary.

M.S. It seems to me that this implies the reinvolvement of elderly people in the normal stream of social activities. There are millions of old people who have been affectively and socially marginalized, hence devalued.

R.D. Yes, I agree with you. But it's difficult to formulate that problem. And on that point I am much more pessimistic than some of my colleagues in the social sciences. I doubt whether the scientific method is of much use in helping us make judgments about social problems. It can enlighten us and enable us to envisage certain consequences, but it can hardly help us to make choices and decisions. The question of integrating older women and men harmoniously into society has more to do with the heart than with reason.

M.S. What are your ideas on the great debate that, along with the atomic question, is the obsession of our time: genetic manipulation, the new terror of the year 2000?

R.D. I have taken a very official stand in the matter by writing an article for the *New York Times* that has been cited in administrative decisions. On the one hand, those manipulations (I was talking only about the introduction of new genes—recombinant DNA) take place spontaneously and constantly in nature. On the other hand, I don't believe in the danger of creating, *in vitro* microbial species capable of setting off epidemics. Obtaining a very virulent new microbe does not suffice to set off an epidemic.

Finally, what we know how to do concerns bacteria and not complicated systems. I would say, somewhat brutally perhaps, that we cannot stop the development of genetic engineering, which technically is relatively simple and inexpensive. Whatever we don't do in the United States or in Europe will be done in Japan or elsewhere.

I must admit that at first I took a position against genetic engineering. After some reflection, I changed my opinion, and in the course of a televised interview I said I was in favor of the continuation of those experiments. Immediately after the interview, James Watson and other colleagues in genetics challenged me to write up and publish the opinion I had expressed orally.

I did so in the *New York Times*. The whole business had a number of repercussions in the academic and medical worlds.

What I had to say had a discernible impact on these people; they were well able to believe in my objectivity in no small measure because I had retired from the world of research. Research in genetics has already produced insulin, somatostin, growth hormones, and interferon, among other things, and all the big industries have set out on this new path. I don't believe one can stop the process, but I would be disturbed if the principle of cloning were applied to organisms more complex than bacteria.

Another subject that has caused me to hesitate for a long time—although there too I have taken a position—is abortion. I was born a Catholic, but I don't believe this has influenced me. I have the scientific conviction that from the moment of conception the new being really exists in the biological and not the theoretical sense of the word, and that abortion thus destroys life. What disturbs me is not so much abortion itself but the possibility that from it we will come, insidiously and progressively, to a scorn for life. But, on the other hand, I also know that abortion has been practiced, and still is being practiced, under disastrous conditions.

I am aware of the deplorable conditions under which certain abortions have been carried out in the United States and elsewhere in the world. That is why, if I were a member of the U.S. Congress, I would vote for the liberalization of abortion laws, even if I am otherwise opposed to that method as contrary to the maintenance of life. But one cannot remain indifferent to the fate of women, and I would unhesitatingly make a political choice against my moral conviction. Abortion is a problem that cannot be resolved with scientific answers alone.

M.S. Are human free will and freedom menaced by the use of the new psychotropic drugs? Is it possible, within the framework of a democratic society, to combat the manipulation of the states of consciousness?

R.D. I can't say that free will is a problem that preoccupies me or disturbs me, and yet I think about it constantly.

I'm going to tell you a little secret. Although I'm French, I've never managed to be truly Cartesian. André Cournand and I are very good friends. Both of us being scientists and New York Frenchmen, it was natural for us to meet and spend time together; and I believe we like and esteem each other a lot. Now it happens that Cournand is much more of a rationalist than I am; and at a certain point, without there being any real difficulties between us, communication stops. I had the same experience with Jacques Monod, whom I knew very well and admired very much. But in certain respects he was a rationalist out of the eighteenth century. Monod always told me, very nicely: "You're only an intuitive type."

In an anthology, I came across some letters written by Descartes to a German princess. In one of these letters he said: ". . . I have come to believe that the most important aspect of health—of mine, in any case—is this: Am I happy or unhappy? If I am not happy, I get sick."

That's very true, but not very Cartesian. . . .

M.S. Can the treatment of mental illnesses really be furthered by alteration of the states of consciousness, either by psychotropic drugs or by electrical means?

R.D. Yes, perhaps, no doubt they can. But, in response to your question, it seems to me important to express to you a deep conviction. No disease has a simple cause. The disturbance of certain biochemical and physiological mechanisms may be the basis of mental illnesses. But these disorders do not necessarily manifest themselves each time a biochemical imbalance occurs. All my personal experience tends to prove that they require special circumstances—a special environment that in the last analysis determines whether a disorder becomes a real, observable disease or, on the contrary, only a potential threat. Allow me to evoke a personal memory.

My first wife was French. Her father had been a worker in the porcelain industry. My wife was in good health. Then in 1942, when we were living in New York, she got tuberculosis and died from it. I asked myself why. I studied her past. Her father had died of tubercular silicosis. She herself, at the age of six, had had tuberculosis but had recovered from it, as one often recovers, spontaneously and with no particular treatment. Then came the war. We didn't suffer from it in America, but my wife was terribly affected by what was going on in France. And her tuberculosis was suddenly reactivated. She was treated in a sanatorium, and came back from there to New York apparently cured. One day as she was passing by Carnegie Hall she, who was a pianist, realized that physically she was no longer able to play. Her tuberculosis was again reactivated, and two months later she was dead.

I have noticed this interaction between the environment and physical and mental health in many cases, including my own. This idea seems to be very banal, and yet it is so tragically true. I have had an ulcer twice. I had the first one in 1921 when, after graduation from an agronomy institute, I couldn't find a job. I had the second one at a time when I was happy but had taken on heavy responsibilities. My ulcer was associated with the fact that I had been given administrative tasks that were deeply repellent to me.

I see the same phenomenon around me all the time. Illness is a refuge—a response to the stress of life.

I have an idea that, basically, mental illness does not differ much in that sense from the tuberculosis I have been talking about. In accordance with the structures of life, there is such a thing as a state of receptivity to illness. This is also true of medication: It is more or less effective depending on the state one is in on the affective plane. We are moving toward an extraordinary evolution in medicine. In my opinion, the great discoveries are not going to come from what we know of cells and biochemistry but from the fact that we will begin to understand the central mechanisms that influence and condition our emotions. What strikes me is that for four or five years a whole series of phenomena has been established in a very precise manner. Take pain, for example. The analgesic effect of acupuncture would seem to be derived from its stimulation of the endogenous production of encephalins in the brain. This is, of course, an empirical observation. But if it is repeated, scientific tools are then employed to explore it thoroughly.

M.S. What do you think about our eventually establishing human habitats in sizeable colonies in space or on the seas as a plausible response to problems of overpopulation or diminished natural resources?

R.D. On this question I have convictions that were originally based solely on practical considerations, but for which I now have a theory. I am now convinced that there are no natural resources.

A resource becomes a resource only after the human mind and science have first recognized the existence of that substance, and then the possibility of using it. Let me give you an example. (There are a great many of them, but this one seems to me the most striking because it is of our time.) The most abundant substance on the earth's surface is aluminum, but bauxite did not become a resource until recently in the human adventure. It was not until the middle of the nineteenth century that we isolated aluminum and perfected techniques for its use.

M.S. Yes. And there is a sentence that recurs several times in your books and articles, one that has struck me very forcibly: "The way in which the landscape is modeled—even the forest, which we believe to be natural—is pure artifact."

So, according to your thesis, most of our so-called natural resources are resources that we have invented and created. But the problem then must be restated in the following terms. Can we ever renew our resources enough—including those that we have created—to provide for a population that is "exploding" in a closed and finite space?

R.D. Once again, the problem is badly stated. I am personally convinced that we could have resources sufficient to feed twenty or thirty billion inhabitants on Earth. The problem is less, "Can we feed more individuals?" than it is, "How do we want to live?"

What quality of life does humanity want? In the absence of a scientific answer, it is interesting to observe the conditions in which our ancestors lived, conditions in which certain aspects of human nature remained as constants.

By "constants" I mean the choice of sites, the orientation of dwellings, the optimum number of inhabitants with respect to the resources of a given valley, the population density necessary for establishing human relations, etc. The valley of the Vézère, in France, presents us with an opportunity to appreciate some of those constants. Although few in number, the men of the Stone Age lived rather close together. Genetically, we have been fashioned in such a way that there is an optimum population density at which people can live together without getting in one another's way. I believe that the present world population is too large, and I'm in favor of limiting it.

M.S. Is it possible that one day immunology will be able to succeed where chemotherapy and surgery have failed?

R.D. Actually, we are only now beginning to understand immunological mechanisms. So that I have every reason to believe that some twenty years from now we shall be able to control immunological reactions much more knowledgeably than we do now. I'm rather optimistic on that point. In the case of cancer, we know that the malignant cell escapes the vigilance of the immunological system, which normally recognizes it and destroys it. But we now glimpse the possibility of recognizing and marking the surface antigens of cancerous cells. Immunology will play an important role in mastering cancer.

M.S. Is not the management of human health care by computer the prelude to an increasingly policelike organization of society, one with no way out for us tomorrow?

R.D. The only hope I have is based on man's exceptional capacity to escape from all controls. Once again, I express my confidence in human nature—in its capacity to resist any will imposed from the outside.

M.S. I will agree with you by quoting a story told me by a very brilliant cybernetician. He imagined the possibility of communicating with other "underground cyberneticians" in a coded language. Every society, whatever its nature, produces its own antidotes to poison and its own ferments of revolt—and here we already see a kind of conspiracy and subversion in the bosom of the cybernetic world.

R.D. That goes along with my optimistic view. . . .

M.S. Will man be able to exercise biological control over his own body by means of miniaturized devices using microprocessors?

R.D. An increasing number of patients in the United States are getting together in groups to help one another learn to live with their disease by profiting from the experience of others. One of the best-known of these "movements" involves diabetics, who have chosen "self-help" as their motto. I have very much favored the creation of these organizations, which allow individuals to take the responsibility for their illnesses both medically and psychologically. But I often remember the saying of the American doctor, Trudeau, who organized the first sanatorium in the United States: "Cure sometimes, comfort often, and console always." And I'll admit that, in reflecting on this, I have become less favorable to these movements. Actually, it seems to me that it is the doctor's job to treat the patient and, in particular, to take on, at least partially, the responsibility for the illness. The word "comfort" carries several meanings: to give aspirin or any other medication alleviating suffering, but also to talk to the patient in a friendly and reassuring way.

M.S. Do you believe that the "new biology" will provide at least partial answers not only to therapeutic problems but also to nutritional, energy, industrial, and other needs in tomorrow's world? In France, we believe that the major advances will come from the United States, which has made more progress in this area than has Europe.

R.D. I am fairly familiar with the problem. For that matter, apropos of the biomass, I'm thinking of the symposium recently organized by a foundation that bears my name, the "René Dubos Forum." Io's an organization in which people from all disciplines get together to discuss a specific problem and to express their points of view freely. For the first symposium that was held, I chose the biomass as a subject (I myself had some light to throw on the subject): "To What Extent Can the Biomass Play a Role in the Energy Future of the United States?"

Of course, as soon as one introduces a subject of that kind in the United States, there is an explosion of curiosity, ideas, and projects. It's extraordinary how enthusiastic and available the Americans are for discussion of such subjects.

I believe that the "new biology" will be able to contribute to the solution of the energy problem, but "contribute" does not mean "solve." For example, thanks to genetic engineering, we will succeed in creating a number of products useful in therapeutics and in the creation of new cells with different properties from those cells we now know.

Allow me once again to bring up a personal memory: the first time that I became well known in the United States. That was in 1935 or 1936. The *Herald Tribune* had organized a huge national competition for electing "the most promising young man of the United States." There were six of us. One was Senator Fulbright, but I've forgotten who the others were. I was the only scientist among the candidates, and the subject of my speech was: "How Can We Succeed, via Biological Manipulations, in Creating Substances Whose Existence We Don't Even Suspect, Particularly in the Realm of Medication?"

I was very optimistic in that respect, and you can see that I haven't changed. At the time, it was a dream, or a futurist project. But today it is reality. In Berkeley there are three industries dealing with genetic engineering. They are still small, but very promising. All the big lumbering companies are engaged in numerous projects. The use of wood residues in Oregon and Washington State has become an important business. And Melvin Calvin's projects for harnessing the solar energy stored by photosynthesis in certain plants in very sunny regions are being realized. In *Fortune*, the *Wall Street Journal*, and all financial circles, people are betting heavily on bio-industry. . . . On the other hand, they are very discreet about what is actually happening. The firms involved are keeping these new operative techniques secret.

M.S. Is there such a thing as preventive medicine that is not coercive?

A Bet on Man

R.D.   I am more and more impressed by the idea that the greatest successes of medicine have been in preventive medicine, which conquered certain infectious diseases, then was directed toward diseases caused by nutritional deficiencies, and is now being applied to industrial diseases. The evolution is recent, but I believe that preventive medicine will open up a vast field of endeavor in the very near future.

Preventive medicine progresses insofar as basic research does. An understanding of the cause of a disease comes from basic research. Today, many treatments focus on the symptoms and not the real cause of a pathological process. We have come back to the debate with which our conversation began. Lewis Thomas, that extremely gifted man, is the most radical representative of the basic research school, which is not at all interested in the medicine of transplants and prosthetic devices—that very precision-oriented and costly branch of medicine. For Thomas, only basic biological science is important in understanding and elucidating the mechanisms of disease. He feels that medical research, in studying the pathological process once it is developed, intervenes too late to understand what operates "upstream."

I tend to agree with him on that last point. Yet, faced with a person who is suffering, we have a duty to use all the tools available to us, even the most ordinary. . . .

ERWIN CHARGAFF
The Beginnings of a New Barbarism

IN HIS ADMIRABLE BOOK, *Heraclitean Fire,* which is both an autobiography and the record of a scientific and philosophical journey, Erwin Chargaff more or less describes himself when he calls one of his chapters "More Foolish and More Wise."

Many dramatic and sometimes droll episodes stand out as landmarks in the story of his life. He was driven out of Vienna, where he was born, by the *Anschluss* and the coming of the Nazis—all this on the heels of Schuschnigg's savage repression of the worker militia of the Social Democratic Party. Like so many others, he came to the United States, where in 1934 he resumed his experiments and began to teach at Mount Sinai Hospital and Columbia University in New York. Beginning in 1949, Chargaff described certain irregularities in the composition of DNA and formulated the concept of "complementarity" (Chargaff's Law). A little later he demonstrated the "pairing of bases," which is the most important proof of DNA's double-helix structure.

But as a biochemist, Chargaff did not "situate" his discovery. The final description of the structure of DNA won Watson and Crick the Nobel Prize in 1953—a prize that had been well earned by the inspired and unlucky Viennese biochemist.

Is Chargaff's gloomy view of the future of science and of humanity due to these misadventures? Or does it stem from his profoundly pessimistic temperament? In any case, his voice thunders in the cushioned, comfortable, and often smug world of the contemporary scientific community like that of a modern Isaiah prophesying the fall of the temple and the arrival of the Four Horsemen of the Apocalypse.

M.S. There are pessimistic utopians and optimistic utopians. I know from having read your books that you are a confirmed pessimist. But are you a utopian?

E.C. Campanella and Thomas More were utopians. . . . None of their predictions came true, because utopians are fantasists. The only ones who are of any interest are the satirists, who are rather pessimistic. That's why Swift is the best of the utopians. He predicted the debacle that has damn well come to pass. *Gulliver's Travels* is one of the rare utopias that has come true, because today we are living among Yahoos.

Let's take the case of cancer research. In my opinion, it's a way researchers have found to make money. The fear of cancer makes the government and public opinion more generous than if it were a question of simply financing basic biological research. As for work on cancer proper, I've seen scarcely any results. I have my doubts as to miracle cures, because cancer remains an enigma—a calamity of which we know neither the origins nor the mechanisms.

Projections of the future are not possible. Everything always happens differently from what was foreseen. Of course, Jules Verne had a few premonitory ideas—submarines and airplanes—but his predictions had to do with technology, which *can* be more or less foreseen. To a certain extent, one can also foresee scientific technology. This means that, if we know the effect of psychotropic drugs, we can imagine the way they'll be used in the future. But I don't believe that one can predict the scientific future in the strict sense. Science and scientific technology are different entities. In any case, the great scientists have always used their imagination rather than their knowledge. Know-how is technique: Methods and procedures can certainly be improved. But one cannot really prophesy scientific developments in the true sense of the term. Innovations are unforeseeable "catastrophes." I don't believe that there has been a single revolution in the natural sciences in my lifetime. The twentieth century did not begin until 1914–1918. World War I marked the beginning of the new era, following a period that was very disagreeable, whatever one may say about the "belle époque." Formerly, the humanities, historical sciences, and natural sciences were matters of individual quest. Between the two wars, I worked in Germany, in America, and sometimes in France at the Institut Pasteur. There were individuals—some of them rather whimsical but gifted, others mediocre and limited—but each of them did his own work and was responsible for his discoveries in a way that has completely disappeared today. The turning point was the arrival of the United States onto the scientific scene in the years just preceding and following World War II. That war changed a lot of things because America intervened brutally and on a massive scale in technology, in the sciences—and finally in everything—in the name of social effectiveness.

I am really inclined to blame the United States for the change that took place at that time. Other countries with vital scientific communities were out of the center of activity because of the war. So the United States had a *de facto* monopoly until 1960–1965. Then the economic crisis worsened; and now we are living through a crisis in the natural sciences, which have become very costly.

Needless to say, it was nuclear energy that gave rise to this phenomenon, which began during World War II with the Manhattan Project. I must say that the Manhattan Project represented the first attempt at an "academic concentration camp" containing thousands of scientists closely supervised by the army. This marked the advent of sizeable teams of researchers working under the direction of an *administrative* authority rather than a properly *scientific* one, and subordinated to the imperatives of a government and its political priorities. Before, scientific work had been an individual and solitary adventure. . . .

The United States epitomizes the twenty-first century in the sciences, which would be what I would call "Alexandrian" by comparison with ancient Alexandria. Today, scientific research costs a great deal, the investments are so great—even in the biological disciplines, which rely heavily on very expensive equipment—that scientific practice has been transformed by it. My generation is coming up against that change, which is not only one of degree but of kind.

M.S. You are definitely not in favor of the large-scale planning of research projects and their distribution among big interdisciplinary teams.

E.C. As I understand it, all of the great discoveries in the natural sciences were made at a time when scientific research was conducted by individuals, with perhaps two exceptions: the discovery of the structure of DNA and genetic engineering. The decline of the natural sciences, as reflected in the decrease of intellectual activity in that arena, has been brought about by specialists who resemble more and more the caste of priests in the Egypt of the pharaohs; that is, their existence is based on their need to survive. This is also true of the great mass of scientists who are creating—secretly, if I may say so—the sciences called "new," because they want to go on making a living and getting grants. In the same way, religion and its rites were invented by priests because that was their *métier;* and by that very fact they destroyed fervor and true piety. Science has been perverted . . . by its tendency to become a regular "business."

Creativity and scientific genius, like mystical inspiration or poetry, are the concerns of an individual and not of a "collective." Scientists constitute a more or less marginal community of individuals—"mavericks," as they say in America. Today, they form a new class that, in order to survive, requires continuity. Naturally, their group has everything to gain by creating problems that make it appear indispensable, and that's what we're seeing now. As you can see, I am a mixture of reactionary and radical. I believe I am rather conservative in my viewpoints, and I persist in saying that most of the problems evoked by your futurologists of health care and those having to do with psychotropic drugs would never have been formulated were it not for this scientific caste.

M.S. Do you believe that stress and depression have been created solely by our current way of life? Weren't there any depressed people in the Middle Ages or the Renaissance?

E.C. I don't know whether people were depressed at the time of the Renaissance. Naturally, there have always been people with problems, but the scientific formulation of those problems did not exist. Psychiatric nosography began to be elaborated about the time of Henri IV, in the late sixteenth century. Of course, some people were not very happy, and others committed suicide; but our age has witnessed a great increase in the number of suicides when compared to the darkest periods of the past. I don't believe that modern man is basically any different from his Neanderthal counterpart, but he has been modified and subjected to new pressures created by the Industrial Revolution, technological progress, and especially the automobile. If you'll bear with me, I think that many of our contemporaries devote the best part of their lives to moving around. They grow dissatisfied and constantly go this way and that—all for nothing, or for very little. This need is recent, if we remember that Napoleon didn't travel any faster than Julius Caesar.

If I were Catholic, I'd think that the devil had damn well taken over the running of this world. With urban civilization becoming more and more demented, I don't see how the man of the twenty-first century could hope to be any happier than people are now.

M.S. Do you see the world to come taking the form of a kind of megalopolis?

E.C. Yes, but the threat of the bomb hangs over us.

M.S. There are energy problems. Perhaps the growth of the megalopolis is no longer possible?

E.C. I believe that the sciences will suffer from a lack of money in ways that we cannot yet foresee. But there will certainly be less energy, hence less production and therefore less money. I have a vision of a new period that will resemble the great migrations described by the poet Claudianus and other writers of the fourth and fifth centuries. I see the beginnings of a new barbarism. The capacity for self-expression, which is a characteristic of man, has dropped considerably both in the United States and in France. To recognize this, one has only to compare the literary production of France thirty years ago with what we have today. But, once again, it is not so much man who is changing as the conditions under which he lives. We are already undergoing an energy crisis, an industrial crisis; and we are living under the threat of the atomic bomb that we feel will explode again one day. History gives us no precedent of a weapon that exists and yet has not been used, except perhaps by mistake.

All of this is naturally the cause of the depression so widespread around us. In the past, from Genghis Khan to Hitler, there were terrible threats to humanity. But the feeling that the end of the species, the total annihilation of humanity, was possible has never before been experienced with such intensity.

M.S. It sounds to me as if you're describing a kind of intuition, felt by the most sensitive people, of some final catastrophe. . . .

E.C. Yes. You know, before every catastrophe, as before an earthquake, there are signs of what is to come. Poetry and music are better indicators of the future than science because they depend much more upon the individual—his mind, his brain. It seems to me that there is a big crisis in the arts, which have almost ceased to exist. There are many people who qualify as musicians and artists, but is there good music or good art today? . . .

The Beginnings of a New Barbarism

M.S. I believe there is.

E.C. I'm not of that opinion. This crisis in artistic creativity is one of the indicators of the general crisis.

M.S. Existential.

E.C. That's a word that no longer exists since existentialism has gone out of style. Today, you have the "new philosophy" in France with Monsieur Bernard-Henry Lévy, who is rediscovering monotheism. In France at least it's all a bit more lighthearted than in America, where people take themselves so seriously. French people have more verve. The Frenchman loves to talk and knows how to express himself; but he too has lost his power of expression, and, with his logomachy and pseudoscientific jargon, he tends to resemble the American. I'm going to surprise you. In the Austrian provinces there are still some writers who seem to be better than elsewhere. For example, in Salzburg and Graz.

M.S. Peasant writers?

E.C. No, not peasant writers. They are, rather, the successors of Kafka. They are very desperate, but they write very well. Bernhard . . . Handke. . . . There are also good writers in East Germany—more than in West Germany. . . .

M.S. Your criteria strike me as surprising but interesting. They parallel your very desperate vision of the world. There is, in fact, a tradition of literary despair in Austria, one that goes back to Kafka, Hofmannsthal, and Musil.

E.C. The Austro-Hungarian Empire encouraged one to get used to the permanent despair of human life. Now, that despair has fled Vienna, which is more than ever the stage set for an operetta, and has found refuge in the provinces. But, to my way of thinking, West Germany is brainless. It has no writers or composers worthy of the name. . . .

M.S. What about Böll, Grass, and others? You are very severe.

E.C. It may be that I focus more on its bankers and traveling salesmen, as in Japan. Bankers who sometimes go bankrupt. I have the curious feeling that France is beginning to resemble West Germany. The Common Market will make all of Europe a colony of the United States.

M.S. Does genetic engineering hold forth the promise of a Golden Age or of an apocalypse?

E.C. I believe the potential usefulness of genetic engineering has been exaggerated. A good deal of interesting basic work will still be made possible by two or three more Nobel Prizes; but I don't believe, personally, that genetic engineering augurs a Golden Age.

M.S. Yet it's the great gamble of science today—the only possibility for the mass production of insulin. . . .

E.C. The pharmaceutical industry will find another way to produce insulin at a lower cost. For that matter, I'm waiting to see if the price of insulin is really going to drop, even if it is produced on an industrial scale.

M.S. You don't think that genetic engineering is going to produce a whole new "rare" pharmacopoeia?

E.C. People are talking about it. But you have to distinguish, you know, between reality and promises. The latter become all the bigger as reality diminishes. I'm very skeptical, but let's wait and see.

M.S. Is it possible to live 120 years? Is it even desirable?

E.C. Where is the limit? Methuselah lived over 900 years, I believe. . . . To live 120 years seems to me possible but not desirable under present-day conditions. If we were living in a Golden Age, perhaps. . . . But right now I don't believe it's a question that will really arise, because in 120 years we'll be living in caves. There will be a ragtag remnant of humanity somewhere in New Zealand that will survive, irradiated and unhappy. . . .

M.S. And yet gerontology and geriatrics are flourishing. . . . There is a whole new pharmacopoeia. . . .

E.C. I do not trust scientists who make great promises. I have little confidence in them. As for the pharmaceutical industry, it will have found one more commercial loophole.

M.S. Can euthanasia, under sociopolitical regulation, be integrated into new morality for tomorrow?

E.C. Certainly! If we don't have the atomic bomb, we'll have to regulate the world's population some other way. But then, is euthanasia really a problem? I doubt it. . . . I think that when it comes to questions of morality and science, my anxiety lies elsewhere—in a certain form of perversion of contemporary science and a rape of nature. . . . There is probably a limit that we should not have crossed over or transgressed beyond, marked by "the two nuclei." One is the atomic nucleus, and the other is the cellular nucleus. One might say that Greek atomic theory—the pre-Socratic atomic theory of Democritus, Lucretius, and Heraclitus—marked a limit for human intelligence. Those limits have been exceeded in my time, starting with World War II, by the fission of the atomic nucleus and the splitting of the cellular nucleus. I belong to the "patient" generation—people who observed and contemplated nature. The scientists who were my predecessors wanted to "know without doing," whereas now our modern scientists want to "do without knowing." Today's scientists are not interested in the attentive contemplation of reality but in changing it. It's a rupture, a truly revolutionary change that has taken place in the relationship between science and nature. . . .

M.S. To you, does that revolution constitute a sacrilege? Are you a believer?

E.C. Yes, I believe it's a sacrilege, a profanation. Am I a believer? Perhaps, in a sense. All scientists are, whether or not they have retained their childhood faith. I would even add that there has not been a single great scientist in the past who was not a believer in one way or another. Because, when one contemplates the beginning of the world, the miracle of life, of its equilibrium and complexity, one necessarily arrives at a form of piety. Lucretius was an atheist, but his atheism was theistic even if he denied the word "God."

M.S. I gather that you are not, like many scientists, a partisan of a biological morality—a morality based on necessity, one that would be better adapted to our time. . . .

E.C. No, I am not. No, there is only one morality. I don't know whether I can define it, but the essence of all religions and philosophies is the same. There are no great differences in the religions revolving around Moses, Buddha, Jesus Christ, and Muhammad except in the forms, the usages, the customs. Our era violates all moralities, all the decalogues of humanity. It is a new barbarism, which tomorrow will be called a "new culture." We are already living in that time. Words have been so debased that today we label as morality what would have been called an absence of morality fifty years ago. Naturally, Nazism was a primitive, brutal, and absurd expression of it. But it was a first draft of the so-called scientific or prescientific morality that is being prepared for us in the radiant future.

M.S. Nazism, though, was not just perverted scientism. . . . It was something else. . . .

E.C. No, it was only that, or essentially that. Nazism was a precursor of the new era, a precursor of the "new scientific morality." Euthanasia and eugenics began under the Third Reich. Before swallowing up dozens of millions of "subhumans," the camps were first built to contain, then liquidate, mental patients, weaklings, etc.

M.S. I do not believe that the present trend in science, even in its divagations and abuses, can be compared to Nazism.

E.C. Fine. My remarks need qualifying. But this new barbarism and its scientific alibi, even without taking the grotesque form of Nazism, is part of the same arrogant behavior of man toward nature. One fact reassures me, however: This cynical, manipulative science will lack the money to achieve its "grand designs." In becoming narrower, under the pretext of specialization, the sciences are becoming so costly in these times of crisis—while the daily life of a citizen in both America and Europe is becoming so difficult—that public opinion will reject it. This is evident already in the United States, where the people, the Congress, and the media are beginning to show a real hostility toward it.

M.S. That's democracy—or what's left of it. . . .

E.C. Democracy, democracy. . . . Let's say, rather, that the safety valve is still functioning in the face of flagrant abuses. No, we have no democracy anymore. Where does it exist? In certain families, perhaps. Look at France. Despite appearances, it seems to me much less democratic now than it was ten years ago. In this respect, France is a portent of the future, but not a very encouraging one. There are many manipulators who are not always very competent. They manipulate and produce nothing. They pretend to act, but they don't really act. And the decision-making is more and more concentrated in the hands of the president and a few technocrats in his entourage. I fear that France is prefiguring the future of political power in the West.

M.S. Should I conclude from what you say that the situation in Eastern Europe strikes you as better? . . .

E.C. I wonder if a society of penury is not more promising for the future than a society of abundance—and even of superabundance—when it exhausts its resources. I foresee not only the "new barbarism" I was just talking about but the material and intellectual exhaustion of the West. We are living in a time of decline. Of course, the phoenix may perhaps rise from its ashes one day; but one has to be much more credulous than I am to hope for it. Take Europe. It is bled white, tired, exhausted. In France, as in West Germany, the only remaining value is cynicism. The countries of Eastern Europe are more backward, and for that reason I have more hope for them. From the few trips I have made in the USSR and the German Democratic Republic, I remember individuals more than the police mentality of the state. Of course, the state, the political system and all that, is abominable. But the men and women I met were more alive, more open. . . .

M.S. Perhaps because they are working in a climate of opposition to the state, and are also poorer.

E.C. They are poorer, but are they really opposed to the existing regime? I believe, rather, that they have become indifferent. They have pulled back from politics and gone into their cocoons. Because of the propaganda, they no longer read the papers, listen to the radio, or watch television. But they have rediscovered a taste for good literature, music, and friendship. That's what I do in New York. Like them, I go into my cocoon.

M.S. To come back to our medical debate, can one imagine that "soft" medicines will some day replace "hard" ones? That immunology, for example, will replace surgery?

E.C. The only medical specialty I still admire is surgery. Surgeons are the only doctors who still have a *métier* and are under the immediate and permanent imperative of failure or success. Their patient will survive, or he will die.

What does one call the health of a people? If we look at the statistics, we naturally see that life expectancy has increased since the sixteenth century, thanks especially to advances in public hygiene. And I exclude the field of public hygiene, too, from my deprecation of modern medicine, because it has allowed people to survive by providing them with drinkable water, sterilization methods, etc. The great hygienists of the past century—the giants of that time, and especially Pasteur—made a considerable contribution. As you know, the grand nineteenth century was extremely optimistic: The Victorians thought that everything was going to become greater, better, richer. Marx was a completely typical Victorian and one of the greatest optimists that ever lived. He believed in the unlimited improvement of humanity, whereas we now see that the future carries with it its own controls and cannot be predicted. According to an asymptotic curve of prosperity and happiness, the nineteenth century raised hopes that were not realized.

Today those dreams have faded away and doctors, for example, practice without either imagination or a sense of responsibility: They have all become "men of science." But the number of true scientists is in my opinion very limited.

In making biological research a "mass" profession, the United States has emasculated the very concept of science. I'm a professional pessimist. All recent progress, including that in immunology, represents success of a scientific kind but not of a practical kind. Apart from antibiotics, there has been little real medical progress. In surgery, progress has been confined to the refinement of its instruments. There have also been advances in diagnostics, such as the scanner, although its scope has been exaggerated. That whole biological revolution that people babble so much about—what has it led to in practical terms, what has it contributed to the care of patients? Where is that "soft" medicine you were telling me about? The split between the basic sciences and therapeutics is becoming more and more pronounced.

The sciences are self-regulating. That is to say, they have created a morality and a universe unto themselves, and live only for themselves. They communicate with one another but not at all with the outside world. Ever since Fleming accidentally discovered penicillin, we no longer talk about decisive progress in therapeutics without laughing.

I know that scientific journalism lives off the idea that we are on the verge of a Golden Age in medicine. But that's a myth. The survival rate of very young infants has improved, and infectious diseases have been partially controlled. Life expectancy has been increased, so they say. I have just reread the *Mémoires* of Saint-Simon. The people he mentions were nobles, and almost all of them lived an average of seventy or eighty years. The exceptions were women, many of whom died with the birth of their first child. Thus, there may have been social progress—but not medical progress; and, apparently, rich people lived better in the age of Louis XIV than they do today. I wonder if our industrial society has not created a "new *mortality*," associated in part with the polluted air that we breathe. There are two ways of surviving: as a statistic, and as an intellect. Our level of intellectual survival is miserable, because most people are waste matter. This decline also takes in medicine in all its forms of practice. I don't believe that medicine will survive as a liberal profession. We see everywhere the tendency toward state-controlled medicine and the reduction of doctors themselves to mere functionaries. We will soon have a profession that is more regulated and probably even more mediocre, meaning that it will be even less able than today to take advantage of "therapeutic breakthroughs," if there ever are any more. . . .

M.S. People hope for so many miracles from the "new biology" that some are already counting on its yielding answers not only to our therapeutic problems but to the nutritional, energy, and other needs of tomorrow. Do you believe that this will be possible?

E.C. Right now, nothing warrants my predicting such a happy future for panbiology. Our future is more likely to be rather humdrum. We are all under the influence of promotional publicity: Everything is exaggerated, and we should forget 90 percent of what we hear. That goes for the sciences as well, since they are, I repeat, a means of survival for the scientists. Their caste has become so big and influential that it has created its own code. As for the biologists' promises that wood, gasoline, and steak will be replaced tomorrow by the products of biogenetic manipulation, allow me to take them with a big grain of salt.

If we eliminate from consideration everything that is not true or proven, not much remains of those golden promises. The great successes of science lie not in medicine or substitutions but in the realm of ideas and our understanding of reality. If one could ask Newton to come back and observe what is going on now, he would perhaps be astounded. I don't believe, though, that he would be very happy, because even Einstein was the precursor of a certain breakdown in the sciences, which had already transgressed beyond the limits that I mentioned earlier.

I don't know if there are any happy scientists today. In the young people with whom I talk a lot I find a terrible lack of tranquility, an evident insecurity—not only monetary or material in nature but, especially, existential. I am struck by the absence of a philosophical point of view in scientists, who seem interested only in proving their hypotheses, publishing their papers, and participating in conferences. Their motivation has become essentially materialistic. I don't want to challenge their morality; but the profession is so much at odds with reality that they have lost all appreciation of philosophical and religious thought. If a scientist wants to be really up to date in his specialty, he must give all his time to it. The amount of knowledge (often marginal) in his field is so inflated that he no longer has the time to read a good book and be a man of culture and reflection. Science today kills happiness—it destroys the well-rounded man.

M.S. Do happiness and scientific progress seem to you incompatible?

E.C. I don't believe that present-day science is bad in itself. It's not the cause of our unhappiness but rather a symptom of it. Here an ontological question arises. Is prolonging life or survival a good thing when one must inevitably die afterward? Naturally, there are always special cases. When the mother of a child survives, it is a great boon for the child. But can one say that the prolongation of life is good in itself? No, not unless it permits a flowering that would not have taken place otherwise. I believe that we are living in a much less happy era than that of Zola.

M.S. Are you familiar with the living conditions of workers in Zola's time?

E.C. Yes, they were very hard. And considerable social progress has been made, but I don't know whether it's really linked to happiness. Actually, human happiness exists only in very limited doses, and I don't believe there is such a thing as unbounded happiness without restraints or limitations. The human condition has remained unchanged since the time of Noah.

M.S. Because of the unavoidable—death?

E.C. No. Even if one were assured of being able to live forever under present-day conditions, all of humanity would commit suicide. It's possible that collective suicides will become a phenomenon of the future—one of the faces of the future if man remains so alienated. It's probable—I haven't seen the statistics—that the number of suicides is already rising in America, in Switzerland, in Sweden, in the so-called happy countries. One form of destabilization that we cannot control has been provoked partially by the modern sciences and in particular by psychoanalysis, which has certainly given us more unhappy people than there were before.

M.S. How many people can still afford the luxury of psychoanalysis? . . .

E.C. In New York, almost all of my students were in analysis while they were in school. It's incredible! I'm speaking of the sixties; perhaps the phenomenon is less widespread today. There were so many disturbed people in the United States that one could easily defend the thesis that the majority of Americans were crazy; only a small number seemed to consider themselves normal. This madness spread to Europe, too. In America, only old people have little or no recourse to analysts. My generation didn't believe in them. When you were unhappy, you didn't go to see a psychiatrist; you tried to master yourself. But my students went straight to a psychiatrist the moment they had to face up to the least trouble. There is usually a psychiatrist attached to a university, or they operate out of an institute all their own. In the big American universities, during the sixties, such "spiritual" help had become a regular institution. Any student who didn't go to see a psychiatrist at one time or another was a bit suspect.

M.S. If I get your meaning, to lean on science in order to understand one's physical or mental state is a bad thing?

E.C. In my opinion, expecting that kind of help from science is impossible, because science gives us only a very partial picture of reality. Very few people realize that, you know. What the biological sciences call "nature" is only a segment, a small fraction of nature. Some 95 percent of nature is not available to the natural sciences, so that biological research involves only a tiny particle of the nature it has isolated. The great majority of problems affecting humanity are not subject to investigation by the natural sciences, which, since the nineteenth century, have defined reality in very limited terms, extending to all of nature only the reality of the little sector they can examine. The great split between humanity and the sciences stems from the fact that the reality experienced by the former is profoundly different from that studied by the latter.

M.S. Then what's left? Mystical intuition?

E.C. Perhaps. The sciences are not in harmony with humanity. They are separated from it. For example, classical philosophers including Kant, Leibniz, Schopenhauer, and Malebranche (I exclude Descartes because he was a prescientist)—tried to understand man in his entirety. That entirety escaped them, and man is no longer defined in terms that correspond to the natural order of things. That great split—the extirpation of philosophy by the sciences—explains why philosophical thought is in such decline today. There are hardly any more philosophers. Instead, we have *déclassé* scientists who are unhappy because they never mastered mathematics. Wittgenstein, who is considered a realist and a positivist, is on the contrary a mystic, as his *Tractatus** makes clear. The split between human reality and the human sciences is deep, I don't see any point of contact between the two. That's why ordinary people no longer understand anything about the sciences. They understand even less than the monks did in the thirteenth and fourteenth centuries. In those days, understanding was inscribed in a scientific universe, limited but approachable, whereas today that universe is unlimited and is relatively comprehensible only to the practitioners of the physical sciences.

M.S. Is humanity going to perish because it has no more *Weltanschauung*? And the churches can, of course, no longer play that role?

E.C. The churches are empty shells, and religions have become social rites. There are certainly very devout Catholics, Protestants, and Jews; but I don't believe they express themselves in the universe of their faith. They are, instead, at the edges of it. Some are even mystics. I have just written somewhere that in America one can find everything, probably even great mystics; but no one knows them because, by definition, a great mystic survives only through his example and writings after he has disappeared. . . .

M.S. Are you addicted to the past?

E.C. No, but I have more faith in the humanity of the past than in the humanity of the future. I am, as you must have noticed, an

---

**Tractatus logico-philosophicus* (Berlin, 1921), the only work of Ludwig Wittgenstein, German philosopher (1889–1951), to be published during his lifetime.

incorrigible pessimist. Most people do not have enough intellectual fortitude to detach themselves from this mad world and live cooped up on a desert island. There are people who still know how to do it, but their days are numbered. And there are very few of them. If we wanted to be prophets of our time, we would preach against the sciences.

M.S. Is it the nearness of the year 2000 that makes you despair so in your vision of humanity's future?

E.C. First, the world does not consist solely of humanity. There are also, in zoos and botanical gardens, many natural forms of life that are going to survive us. I don't foresee the end of humanity, because even the atomic bomb will not be able to annihilate us completely. I am not optimistic as regards present-day man because, as I have repeatedly told you, I believe that the natural sciences are a tool of his degradation. The scientists of the past—of the nineteenth century and before—were, above all, realists and hence did not practice the idolatry of the sciences. Human intelligence made extraordinary leaps in that period, but those scientists were still counted among the philosophers. The natural sciences actually constituted a branch of philosophy and were a means of understanding the world. There is a great difference between understanding and explaining: Understanding is much more basic than explaining. In fact, one can explain much more than one can understand. Today the sciences have become exclusively explanatory. That is, they place themselves on a level that is broader but at the same time more superficial. I admit that I belong to a generation that tried harder to understand than to explain.

M.S. But everything you are denouncing—atomic fission, genetic engineering—is already with us. We can't put it back in the box. It's out.

E.C. Yes, it's probably the work of the devil. I told you that earlier.

M.S. Have we defied God or nature?

E.C. I don't know whether or not we have defied nature, but we've altered it. We've literally denatured it. We've created an imbalance that our intellectual and moral tools are not capable of straightening out or mastering. Since World War II, there has been a fissure between our moral capacities and our intellectual capacities.

I'm seventy-four now, and I belong to a generation you might call "marginalized," meaning that nothing remains but memory. I realize that, materially, people are living much better now than in my youth. But there used to be a certain continuity between generations, and this

The Beginnings of a New Barbarism

has disappeared. Actually, we probably had, right in front of our eyes, signs of the times to come. One always feels one is in touch with the past, even when already cut off from it. That's probably what the newborn feels when the umbilical cord is cut: He is still suspended, although the cord no longer exists. Thus, the big cutting off, the great gulf between the past and the future, didn't become perceptible to people until World War II. Although I belong to the past, I see a lot of young people with whom I communicate easily. It seems I'm very popular with my students. They understand what I have to say better than most of my own colleagues do. Perhaps they perceive, as I do, that what is in store for us is not a rose garden. No one knows what the future will bring. But I take a dim view of things because I don't see a single encouraging sign: The *modus vivendi* of the future seems to be genetic monstrosities, the atom, euthanasia, prosthetic devices. . . . Man as merchandise and "merchandized" will become waste matter much sooner than the commodities themselves. He will "spoil" much more easily, because he is more than his "material" self. The material self is surpassed by what one might call the soul or the spirit. Today there are no more prophets, philosophers, poets. The arts and humanism were buried with the dead of World War II. Picasso, for example, was the sole survivor of an earlier age. When I saw the Picasso exhibit at the Centre Pompidou [Beaubourg] and compared his work with that of his successors, I felt a break that I couldn't define but that is evident. Our time will be marked by progress in physics, the fission of the atom, and in the same stride by the demise of a race due to Hitlerian genocide. Hitler was a key to our time; he was a precursor of our sciences. In the not too distant future, monuments will be erected to him. The birth of the New Right in France supports this view.

M.S.   A university president in California told me he wanted to put up a monument to Hitler, but not for the same reasons. By driving out very talented scientists like you, Hitler enabled his university to become one of the most important in the world. . . .

E.C.   I'm not a refugee, I'm a nomad. I was in America in 1928. Then I went back to Europe. I was in France in 1933 and 1934, and I again came back to America. I am not a typical *émigré* of that period. Rather, I'm a sign of bad times—a bird of ill omen. Among the *émigré* scientists on that list you allude to, I don't know if a single one would understand me. The young understand me better; and you can't imagine the despair evinced by biology students in the United States. They have an insatiable hunger and thirst for something they can't define. It isn't ordinary prophets they need but reasons for living and a little rationality amidst all our present-day chaos. Diderot is making a big impression right now. The materialists of the eighteenth century could be our prophets and could occupy a position very different from what they did in their own time. We no longer know where we're going, and we no longer know what we're doing, because the natural sciences contain an element of unreality that looms large when one works in them for too long. If you start trading in human ears, after ten years a routine will be established, and you'll no longer realize that those are really human ears you're buying and selling.

M.S. What do you say to the young biochemists and doctors you teach?

E.C. That they should change professions! I depict present-day reality for them in my own words, which are naturally exaggerated and apocalyptic. Einstein said toward the end of his life that if he had it to do over again he would be a locksmith or a gardener. . . . When you are faced with a monster you can't combat and hence overcome, the only possible recourse is flight.

M.S. When you discovered the structure of DNA did you know what you were doing? Were you aware that you would be to some extent, and even very much, involved in the early stages of development of what you consider a demon: genetic engineering?

E.C. No. This happened between 1947 and 1952. I'm one of those prophets who should curse themselves for the evil they've done without knowing it. I'm a chemist and was interested in solving a problem. I began my work under the influence of Avery. I knew that DNA contained, in a form as yet not determined, the principle of cellular specificity. But I didn't foresee what would happen; I was too isolated, too wrapped up in my own problems, which had more to do with the philosophy of nature than with its structure. I approached chemistry as a philosopher rather than as a scientist.

M.S. You knew that your work would involve the essence of living matter. Didn't that bother you?

E.C. No, because I've always made a distinction between understanding nature and explaining it: Explanation is much less important and much simpler to arrive at. I wanted to understand what would be good for the mind, for the human brain, because understanding nature seemed to me to be a good thing. Explaining, or utilizing, nature involves ambiguity. The snake in Genesis was not of my opinion when he said: "Ye shall be as gods, knowing good and evil." I thought that

The Beginnings of a New Barbarism

to be able to understand nature was good, but that to be able to use it was of dubious value. Consequently, I never would have studied the problems that are now basic to genetic engineering. I'm only sorry that this interview didn't take place twenty years ago. . . .

ANDRÉ LWOFF
Passion and Reason

WITH FRANCOIS JACOB and the late, lamented Jacques Monod, André Lwoff has been a member of that astounding trio of scientists and personal friends who have brought three Nobel Prizes to France and guaranteed new life to the venerable Institut Pasteur. Their great adventure in science, and in friendship, will probably make a lasting mark on the intellectual history of France.

A biologist and virologist of almost eighty, André Lwoff pursues a triple career as scientist, artist, and public figure. His tall, lanky silhouette still haunts the corridors of the modest skyscraper that the department of molecular biology erected, rather incongruously, in the dusty part of the rue du Docteur-Roux.

A painter of the Catalan countryside where he spends several months of every year in his country home amid the vineyards and scrub brush of Banyuls-sur-Mer, Lwoff gives us a brightly colored and sensuous *oeuvre* that seems to belie the apparent austerity of the man himself. But it is especially the public personage, the passionate defender of

Passion and Reason   121

human rights, who attracts the attention of the media and of public opinion. Is he an inspired prophet or a naïve Don Quixote gone astray on the perilous paths of *Realpolitik*?

A man of reason and courage, Lwoff defends dissidents from the Soviet Union or Argentina, persecuted minorities, and especially Russian Jews who want to emigrate. He defends them with a cold passion and a vehemence that spring from a remarkable determination and constancy in this moderate man of science.

Faced with the "Finlandization" of Europe, with this new "treason of the clerks"* that each passing day gives more away, sometimes to the blackmail of the oil cartels, sometimes to the complaisance with which leftist vogues are accepted, Lwoff has consistently chosen the "unpopular side": Israel, the defense of democracy everywhere in the world, the defense of liberties in France, and international cooperation.

He has lots of friends and lots of enemies, but that was also true of Galileo, Erasmus, and Einstein.

M.S. I've often wondered if utopianism wasn't one of the principal elements in scientific creativity: imagination coming to the aid of intuition, the passion for seeking and projecting one's research toward all possible futures. But, on the other hand, isn't utopianism in a biologist or a doctor dangerous in that it raises false hopes? Intellectual disappointment for the man of science, despair for patients with cancer or disseminated (multiple) sclerosis?

A.L. Every great discovery at first seems utopian. The prevention of infectious diseases is one example. Smallpox vaccination was the first real success of medicine. Pasteur put the prevention of infectious diseases on a scientific basis, and now we can protect people against many bacterial and viral diseases. It is not utopian to think that new vaccines will be discovered. Another great triumph of medicine has been the *treatment* of infectious diseases. It is obvious that chemotherapy and antibiotic therapy are going to be further developed. The treatment of viral ailments is now well under way.

One can easily imagine breakthroughs in the therapeutics of hormonal disorders and disturbed mental functions. Unfortunately it is not so easy to imagine our learning to prevent two of the most common causes of disease and death: addiction to alcohol and addiction to tobacco.

Utopianism is not dangerous in itself. What is dangerous, in the realm

*The reference is to *La Trahison des Clercs*, a book by the French critic Julien Benda dealing with the attitude of the intellectuals (*"les clercs"*) during World War I. [Tr.]

of science as in that of medicine, is skepticism, which is debilitating.

M.S. Of the so-called drugs of the future, and specifically from among the psychotropic drugs (see pp. xvi–xvii), which ones strike you as being both plausible and promising in their use?

A.L. I'm very suspicious of the panoply of psychotropic drugs. Certain drugs used to relieve insomnia engender serious lesions of the hematopoietic system. The insomniacs who use these drugs get to the point where they can't do without them. Drug dependence and actual drug addiction are thus created. It is possible that some of the new psychotropic drugs will cause new kinds of dependence.

As for the hope of one day discovering compounds that will provoke "a new awareness of the beautiful," this strikes me as totally unfounded.

M.S. In any case, some of these behavior-conditioning psychotropic drugs are already being used, in totalitarian societies especially. In the Soviet Union, it is current practice to use psychotropic drugs on political prisoners and dissidents.

A.L. In such cases, we are, of course, no longer talking about the practice of medicine. The treatment to which dissidents confined in special psychiatric hospitals are subjected aims at destroying their personality and their willpower. Therapeutics no longer figures in any of this. For that matter, the future of health—and I mean moral and mental health as well—has more to do with politics than with medicine. In totalitarian countries, as you know, there is no freedom. There is neither freedom of expression, nor the freedom to practice a religion, nor even the freedom to choose one's reading matter, since many books are banned. This situation—stereotyped behavior's being imposed on an entire population—is extremely grave. To speak of the progress medicine has made is really a joke.

M.S. Yes, putting psychotropic drugs to such use is particularly equivocal, if not revolting. But, in a general way, psychotropic drugs—as is true of all medication—satisfy a market demand. One may deplore this, but it's a fact. Like Prudhomme's sword, these drugs can either be used to defend the republic, or they can be used to cleave it in two. In other words, some can at times be used to manipulative ends, as in the USSR, while others—or the same ones—can be used positively, as psychological prosthetic devices. In what ways can they be used conscientiously?

A.L. It seems to me very dangerous, even monstrous, to base the intellectual and moral formation—the equilibrium—of the human personality on the use of drugs. I feel that in a society, of whatever kind, the aim of education is to help individuals live in harmony with their fellowmen and their environment. What I am saying is, however, completely utopian if we bear in mind that fanatical notions and groups of all kinds, both religious and political, are resurfacing in a growing and disturbing way.

M.S. Certainly. But in a democratic society nothing stops people from consuming drugs, alcohol, and tobacco. Is there such a thing as democratic regulation of psychotropic drugs?

A.L. By definition, regulation cannot be democratic. The first duty of government is to create conditions of life and work favorable to general equilibrium and to people's adaptation to their milieu. It is undeniable that living conditions in the big cities are unacceptable and generate psychic disorders. People who must spend several hours a day in mass-transport systems to get to work, and who live in dormitory-cities, are leading inhuman existences. Are they free, even in our democratic society? I don't think so. Even if they have enough to eat, they have no real freedom—the kind of freedom that would allow them to have leisure time, to cultivate themselves, to engage in activities outside their work. One must first of all combat such urban realities.

M.S. Hence, recourse to those crutches called psychotropic drugs.

A.L. Yes, but they're a bad solution. Not even a solution but a mediocre palliative. It would be better to have no such drugs at all. . . .

M.S. Let's consider twenty-first century man. Do you see him as more peaceable, more congenial? Or will the population explosion and its consequences—the increased density of the urban fabric, the increased scarcity of natural resources—make him even more aggressive?

A.L. The increase in violence, the spread of totalitarian regimes, the upswing in political and religious fanaticism, do not allow for a very optimistic view of the evolution of man and societies. Culture, humanism, and Western civilization in general are in danger of dying. The increase in the crime rate, especially among the young people in our overpopulated cities, is a grave sign.

M.S. Is today's aggressiveness the expression of a void—of an absolute lack of meaning in life?

A.L. Religion played a very important role in the equilibrium of former societies: in France, for example, until the nineteenth century. But it did not prevent wars of religion, massacres, or the inhuman and aberrant behavior of families—and particularly of mothers of families—

toward their children. I refer to Elisabeth Badinter's remarkable book.* In the seventeenth and eighteenth centuries, nursing babies were sent off by their parents—they didn't even know where—under such horrific conditions that many of them died en route. Families very often did not concern themselves with their children until they reached school age. This is not aggression per se, but, in the final analysis, indifference is a negative and yet very real form of aggression.

M.S. In our time, violence has taken an anarchical, scattered, and almost anonymous form. It is the doing of men acting alone, of terrorists fighting the system; that is, fighting against everyone and even against themselves. Strictly speaking, ideology does not exist. It has been reduced to a few simplistic slogans from which the notion of a utopia is excluded.

A.L. In a sense, you're right. It's a violence that is not based on a religious faith. In a way, it's a violence born of solitude and despair. But I don't agree with you as to its lacking a political aim. The political aim is the destruction of the capitalist system. We've chosen an extreme case. But I persist in believing that the answer to violence and dissatisfaction with life is not a matter for pharmacology to resolve.

M.S. Does genetic engineering promise us a Golden Age, or does it harken an apocalypse?

A.L. Neither one. The transformation of man by genetic engineering lies in the realm of science fiction. For the time being, genetic engineering has been used mainly for making substances that were difficult to obtain by means that might be called natural. A hormone that could not be extracted from a gland can now be made using bacteria. In spite of much groping and some failures, remarkable results have been obtained regarding insulin, and the growth hormone. It is now hoped that human interferon can be produced by genetic engineering. Thus, a certain number of drugs will be more accessible and less expensive. This technology will no doubt also make it possible to develop, perhaps in a spectacular way in certain cases, the manufacture of vaccines. For example, the antigen of a hepatitis B virus has just been produced at the Institut Pasteur. This is an important step toward making a vaccine against viral hepatitis. But all this progress does and will not suffice to bring about a Golden Age in therapeutics. To expect that much from this kind of technology—to hope to change man—is chimerical.

M.S. Is it possible, or even desirable, to live 120 years?

---

*Elisabeth Badinter, *Mother Love*. New York: Macmillan, 1981.

A.L. It might be desirable to live 120 years if a certain physical and intellectual decline did not set in long before one reaches that age. Longevity is partially governed by genetic factors. Excesses of all kinds can diminsh it, while a balanced life can prolong it to a large extent.

M.S. You have lived a long life. I admire your sprightliness and your creativity as a scientist, as a public personage, and even as a talented painter—to the extent that I can judge such matters.

A.L. Let's not exaggerate: I'm not a hundred years old. I like equilibrium. I neither smoke nor drink. Those drugs create a dependence, which I view as a decline, a degradation. Without being a puritan, man should learn to master his passions, at least some of them.

M.S. Can euthanasia, under sociopolitical constraints, be integrated into a new moral code?

A.L. Totalitarian "morality" subordinates respect for the individual to the interests of the state. To answer your question about euthanasia, we must distinguish clearly between death dealt out to alleviate suffering and death dealt out to eliminate individuals considered to be socially useless, dangerous, or "inferior." Only the first kind of euthanasia is acceptable. In certain specific cases—extreme cases—it's up to the doctor to decide. But the doctor does not feel free, and he is *not* free, because it's his duty to save the patient and to prolong life whenever possible. The doctor lives in a sociopolitical, cultural environment from which he cannot always remain aloof. Some Catholic doctors will never perform an abortion, even if the mother is in danger or she already has a large family living under miserable conditions. Those same doctors will not perform an abortion if a girl has been raped by her father or her brother—something that happens from time to time. . . . No, the doctor is not entirely free in his actions: He is conditioned by his professional duty, his personal morality, and by his religious faith: all imperatives he obeys. It seems to me desirable, however, to help a patient die when he expresses the wish to have his suffering cut short. Some doctors do this, but they don't admit to it. An English gynecologist told me one day that in the course of his lifetime he had ended the lives of a number of newborns with monstrous deformities.

M.S. Without getting into trouble with the law in his country?

A.L. He said this much later, when he was no longer practicing. I don't believe he had any trouble. There are things the doctor must do without saying he is doing them. If a doctor sees a monster born, it is very easy for him to see to it that it doesn't live. It is the doctor alone who should and can take that responsibility, to spare the child a life of suffering and to spare its parents martyrdom. Parents who have a drastically abnormal child are doomed to a very difficult and painful life. It is impossible to ask the mother to make such a decision. And no legislation can intervene in situations of that kind.

M.S. Can't there be—among doctors, among men of science—if not precise rules then a kind of general directive specifying certain limits that must not be overstepped?

A.L. That would be very difficult, if not impossible. Each case is unique. Let's suppose that the parents ask a doctor to deny life to a child who is plainly monstrous, who will have an empty existence, who will impose a terrible life on his family and will be a burden to society. If euthanasia were a violation of the doctor's convictions, he would not comply with the parents' request. It's a question of education, too. I believe that the chief of service should discuss such cases with his students, his colleagues, his assistants, and suggest rules of behavior. To legislate in this realm is, however, inconceivable.

M.S. Basically, then, you're saying that every hospital and every clinic should have a kind of committee of wise men, a bioethical council in which the chiefs of service and their assistants would make a decision that would be specific to the hospital and at the same time relatively general in application within that context. The fact that the doctors in that hospital would then be acting in concert would make it possible for them to arrive at a loosely prescribed course of conduct.

A.L. Yes. It's a question of moral professional training. Many chiefs of service give this to their students, because they consider it their duty to teach not only clinical medicine and biochemistry but moral behavior.

M.S. What do you think of your American colleagues (in California, among other places) who carry some sort of fireproof card asking that, in case they are in a desperate situation, a gentle euthanasia be practiced on them, to be followed by the donating of organs, cremation, etc.? In other words, that card constitutes a biological will.

A.L. I find their attitude worthy and respectable.

M.S. Does man have reason to fear that his free will, his freedom, will be jeopardized by the new psychotropic drugs? Can we, within the framework of a democratic society, protect ourselves against excesses of that nature, which involve the manipulation of the psyche? And if so, how?

A.L. Consciousness is already sufficiently manipulated without psychotropic drugs or other such methods. The mass media modify consciousness and represent a powerful and dangerous weapon in the hands of political parties and organized religion. To protect the citizenry, it would be necessary to suppress radio, television, and the newspapers, and teach man to think for himself—which is, of course, impossible.... But let us not speak ill of *all* mind-altering drugs. The trouble is that under the same generic heading of "psychotropic drugs" are grouped those that have damaging effects on the healthy person (those that you called psychological prosthetic devices earlier) and those that have a beneficial effect on the mentally ill. And thus causes people great confusion. Certainly we have much to hope for and expect from drugs that modify the physiology of the nerve cell. Quite a few mental illnesses are associated with biochemical disorders, and one can hope that in the not too far distant future some of them will respond to drug treatments. I don't believe that electroshock and related therapeutic approaches can intervene in a valid way; but treatment with medications may yield rather spectacular advances.

In the past we have resorted to confinement, barbarous treatments like electroshock, and symptomatic treatments like bromide and valerian, but nothing else. The patient was brutalized. Today we strive for a specific action on the function or functions responsible for the abnormal behavior in question. Moreover, our knowledge of the physiology of the nervous system is increasing. The discovery of new "psychotropic" substances will thus be useful in the treatment of various mental illnesses.

M.S. What will sexuality be like in a future world where, from the contraceptive vaccine to the test-tube baby, conception and sexuality will be totally dissociated?

A.L. Society has become more and more permissive, and the young are liberated—or so it seems. Young people are more and more sexually precocious. Is that a good thing? I can't say, since I'm not a moralist. But as a physician I can state that this sexual freedom is one source of a renewed outbreak of veneral disease. In the United States it is common for young people of fifteen to set up housekeeping, to live together for six or seven years, and then to get married. From what I know of it, the experiment is not conclusive, since the fact of living cooped up like that for several years when one is so young does not favor the couple's equilibrium. I might add that quite a few young people are virgins when they marry, but no one talks about them.

M.S. Can you imagine a world in which, thanks to prosthetic devices and transplants, defective organs will be replaced as routinely as the parts of an automobile? What ethic could be devised to cover the establishment and administration of "organ banks"?

A.L. As for prosthetic devices, I cannot conceive that any of them, once implanted, would function like "natural" organs. Then, too, we must not forget the economic aspect of the question. Right now, prosthetic devices and organ transplants are the privilege of a very small fraction of patients, and they will remain such for a long time to come. The cost of this kind of care increases considerably as therapeutics becomes more and more sophisticated. Society is being confronted with a choice. Should we unreservedly spend enormous sums to save, or permit the survival of, a single individual? Or might we do better to improve the lot of several hundred others?

M.S. In a certain number of countries, people have opted for preventive medicine—social medicine—rather than high-tech medicine. But doesn't that mean giving up the advances made by research in high-tech medicine? And I consider this important because, in the last analysis, that research should some day be beneficial to medical practice as a whole. Allow me to make this rather trivial comparison: If, in the matter of the manufacture of automobiles, you give up competition, which is very costly, don't you also give up the possibility of improving the ordinary car?

A.L. I don't think so. In any case, we will never stop the researchers from searching and making discoveries, even if—for purely economic reasons—their research bears no real fruit. I realize that the study of transplants has made for considerable progress in immunology—progress from which, one day, the whole world will benefit. The difficulties inherent in heart transplants have been solved technically, but this is only a temporary solution. People who have undergone heart transplants have to take immunosuppressive drugs. This problem has not been entirely solved, and may never be. Since all individuals are different, a donor's having a heart identical to that of the recipient is a categorical impossibility.

M.S. Are we at an impasse?

A.L. Yes and no. Some patients have received kidney transplants and been saved. But these patients must necessarily undergo immunosuppressive treatment. Now (and this is a fact that isn't much talked about), the incidence of cancer among individuals subjected to that kind of treatment is higher than it is throughout the general population. This phenomenon is understandable, because the immune defenses play a primary role in the organism's resistance to malignant cells. A treatment that suppresses that system of defense permits the development of certain malignant cells that will subsequently form a tumor.

M.S. Will immunology be able to take over where chemotherapy and surgery have failed?

A.L. It is clear that immunology will not be able to resolve all problems, but it is also clear that immunology will be used increasingly and will make for spectacular progress in the field of therapeutics. Let's take the example of interferon, that protein normally manufactured by the organism that plays a role in combating viral infections, and perhaps also in the elimination of certain tumors. Today much research is being conducted to find a way to produce interferon by genetic engineering. The injection of interferon into a patient constitutes one example of immunotherapy. There are others, too. I'm thinking also of those people who for genetic reasons have a deficient immunological system. An immunological approach consists in using adjuvants that facilitate the production of antibodies. One of the best adjuvants is BCG; BCG's active substance, identified by Edgar Lederer, makes it possible to confer antigenic power upon substances deprived of it. . . .

M.S. Is the management of human health-care by computer the prelude to even more policelike social organization, leaving us no way of escape in tomorrow's world?

A.L. I don't think so. The fact that one has computer data on the health of a person has nothing to do with his political or extrapolitical activities. They are two totally different realms. . . . With this we are leaving the purely biological problem behind and entering into that of survival under a totalitarian government. If the latter decides to end the lives of all people who are circumcised, or who have red hair, or who don't "think correctly," there is no need for a computer printout.

The medical file already exists for many people, since it is compiled at the time of military service. It is not a genetic file, strictly speaking, but it is not very far from it. If you have a genetic deficiency, they know it, or they will ultimately know it. We should not take too dim a view of data processing. In matters of health care or elsewhere, it will be good or bad depending on how it is used.

M.S. Will man be able to exercise biological control over his own body by means of miniaturized appliances containing microprocessors? Is this desirable?

A.L. Man is so constituted that a mere apparatus of some sort cannot help him. Some people eat too much, but some little device or other will not stop them. Not, that is, unless you put a ring around their necks like the ring fishermen use on cormorants to keep them from swallowing the fish they catch. I repeat that, no matter how perfected a device may be, it cannot replace personal discipline.

M.S. The Americans have developed an amusing gadget. It's a cigarette dispenser, sealed like a strongbox, that dispenses a single cigarette every hour. Actually, if the smoker has enough willpower to wait an hour, he should have enough not to smoke at all. But in the case of the diabetic, it's not a question of willpower. I'm surprised that you take such a radical and negative attitude toward devices that might, for example, control the blood-sugar level, and which would no doubt sell by the millions.

A.L. That quantitative argument doesn't mean a thing to me. Cigarettes are sold by the millions, but that doesn't stop their being bad for one's health. The fact that a gadget sells in large quantities does not prove that it is useful.

M.S. You are probably a man endowed with a lot of willpower, but you have a very harsh and austere view of life.

A.L. No, I'm not at all austere. I've always regarded research as a game. And the same goes for writing, painting, and many other things. I've played all my life. I'm not a puritan. What strikes you as puritanical in me is in fact clearheadedness.

M.S. People hope for so many miracles from the "new biology" that some are really convinced it will provide answers not only to our therapeutic problems but to our future nutritional, energy, industrial, and other needs. Do you believe this to be true?

A.L. Bio-industry has existed for a long time, but it has not always been called bio-industry. The selection of vegetable and animal species useful to man has been actively practiced for several millennia.

The "new biology" cannot solve all our energy problems. In industrial socities, the consumption of energy is considerable and is increasing rapidly. In the future, energy will be provided first by nuclear sources, then by solar, geothermic, and aeolian (wind) power. . . . Several aeolian devices have been successfully developed. As for geothermic energy, it is off to a good start now, as you know. And the same goes for the use of solar energy.

It is finally possible to recover—thanks to microbes—cellulose waste matter that is not now being utilized. This could constitute a partial answer to energy and nutritional needs. But the recovery process is not total. The exhaustion of raw materials is no doubt a great threat to humanity. When there is no more iron, copper, nickel, or zinc, a new life-style will impose itself on us. Industrial Western civilization represents a tremendous waste of materials, mineral elements, and energy as well. This cannot continue. I'm convinced that the energy problem will be solved, but I don't see any solution for the problem of the exhaustion of raw materials. Biological research will not be of much help there.

M.S. How do you see the role of the doctor and "medical power" evolving in the future?

A.L. The doctor's role is to care for sick people, and it will go on being just that. As for "medical power," it has so far been incapable of eradicating alcoholism and the addiction to tobacco; and I do not foresee its being able to check the development of the main causes of intellectual and emotional retardation: television and animated cartoons.

M.S. Can you conceive of a preventive medicine that is not coercive?

A.L. To make compulsory a vaccination that protects the individual and society is a constraint, just as the banning of narcotics is. Some constraints are not generally accepted. Others are inconceivable: for example, banning alcohol and tobacco. But, in any case, life in society implies the imposition of constraints—necessarily.

In my opinion, persuasion is better than force, or authoritarianism. I believe that the best example is vaccination against diphtheria. It was compulsory in France; and a study made a few years ago showed a large number of young people had falsified vaccination certificates. Their proportion was especially high among children from military officers' families. And that's perfectly understandable. The colonel tells the military doctor: "Doctor, make out a vaccination certificate for my son." "Yes, Colonel." Thus, in France, despite the legal obligation, almost 50 percent of the population was not vaccinated. In England diphtheria shots are not compulsory, but a big publicity campaign was launched: distribution of documents, radio appeals. . . . Thus, the English became convinced that it was their duty to protect the health of their children.

Today, in Great Britain, virtually the entire population is vaccinated. And the same method was employed for the BCG vaccine. In France, too, "incitement" has proven by far preferable to force. As long as the BCG vaccination was not made compulsory, it was accepted by the bulk of the population, which was well informed and motivated. But, from the day the vaccination became compulsory, the authorities ceased their public campaigns, and the number of false certificates has mounted.

M.S. In Canada, I was struck by a remarkable development in all kinds of associations and experiments concerned with sociomedical innovation, the goal of which was to teach individuals to "self-manage" their health. I must admit that for me the word "self-management" seemed a bit demagogic and nonsensical. Then a few Canadian friends invited me to sit in on consultations at some of the well-known local centers of community services, located in low-income areas. I was surprised to discover informational posters, brochures and magazines, even hostesses. And the people running those centers—and this is the most important thing—are really concerned with their clients and their requests. I know that interesting results have been obtained from taking into account the individuals' own willingness to take responsibility for their own health.

A.L. Certainly. But it must also be said that the Anglo-Saxon countries are basically different from France. Even in French-speaking Canada, there are Anglo-Saxon character traits; the *Québécois* do not behave like the French, if they'll pardon my saying so. They have more civic pride. They act much more as "concerned citizens," conscious of their duties and responsibilities toward their own health.

Education and persuasion are better than force. One must also take into account the mentality of the population being addressed. "Self-management" of one's health would mean precious little to members of Jehovah's Witnesses to, who refuse all human intervention in the course of a disease.

M.S. In the Western world, at least, it has never before been so easy to gain access to certain kinds of culture and communication. Never has public hygiene been so widespread. You can eat safely in the most miserable joint. You can drink water from any old faucet. Consequently, the objective conditions for happiness and good health are present. Never have there been so many inexpensive leisure-time activities—libraries, shows. The sociopolitical network is stable; and, despite the current high rates of unemployment, a relative prosperity exists.

A.L. And yet the incidence of mental illness is increasing. . . . Psychic disorders are multiplying because of tension and anxiety about the future, not to mention noise and other nuisances and kinds of pollution. . . .

M.S. Aren't you evincing the brand of naïveté derived from an addiction to the past, that certain ecologists share? In the past there was at least as much noise, if not more. There were discomforts and pollution of all kinds. Think of the inconveniences of life in Paris as described by the chroniclers of the eighteenth century. When people dumped their garbage and emptied their chamber pots into the street, was organic pollution less serious than chemical pollution is today?

A.L. It wasn't any *more* serious, in any case, since longevity *did* increase. If, in the West, most people have enough to eat, there are still many who know poverty. Today the unemployed heads of households that live on their unemployment compensation are existing under conditions not calculated to improve physical and mental health.

M.S.   Certainly. But fifty years ago they would have died of hunger, quite literally. I believe the present malaise grows from something else, perhaps the increased scarcity of things that were once believed to be in inexhaustible supply. For example, in certain Western European countries living space is considerably reduced. People will be living more and more on top of one another.

A.L.   If we can't remake the world, we can try to rearrange it. I'm thinking of an example of "social rearrangement" practiced in Germany before the war. The Germans had decided to transfer from the big cities to the countryside a good number of factories making spare parts. In this way, the workers gained very suitable living conditions. The experiment was a success. In social matters—and I include health care—what we lack above all is imagination and courage.

GABRIEL G. NAHAS
Pleasure and Dependence

STRADDLED between New York and Paris, between Columbia University's medical school, the UN (where he is one of the most respected experts on the Narcotics Commission), and the cellular toxology department at the Fermand-Vidal Hospital in Paris (INSERM), Gabriel Nahas, a biochemist and pharmacologist, has made the fight against drugs into a personal crusade—a crusade that he wages with stubborn vigor on both sides of the Atlantic.

For this Franco-American professor, age sixty, there are neither soft drugs nor hard drugs, just one and the same scourge destroying the willpower and personality of individuals subjugated to the tyranny of immediate pleasure via the addict's dependence on drugs.

Gabriel Nahas sees himself, above all, as a man of the laboratory. For the past ten years or so, he has oriented his research toward the biological effects of psychotropic drugs, and in particular of the derivatives of cannabis, of which THC is the active substance. His work would not appear to have convinced everybody.

Pleasure and Dependence

His theses and the glaring light he projects on our "diseases of civilization" through his experience as a clinician and toxologist, although they do not meet with unanimous agreement in the scientific community, have isolated the drug problem from the nauseating romantic mythology that has so often made drug use seem colorfully attractive.

It matters little whether Gabriel Nahas's crusade is that of a Puritan trying to confer scientific legitimacy upon *a priori* convictions or whether, on the contrary, it is that of a rigorous scientist drawing his deepest motivations from his laboratory work.

M.S. We are living in a society approaching a millennium. We are all waiting for the year 2000, and predictions are coming thick and fast. Some are optimistic and even euphoric: Hermann Kahn, for example, thinks man will manage to overcome all obstacles despite the scarcities he faces. Others, like those of the famous Club of Rome, take the apocalyptic view. As a man of science, what is your view?

G.N. Since I am a biologist, my view is extremely colored by what I know of human biology: its potential, of course, and also its limits, which are apparent to all who are now studying what man is becoming. I believe, in general, that the futurologists have not paid enough attention to biology's limits and have let themselves be carried away by utopian fantasies. On the other hand, the economists—you mentioned the Club of Rome, and I'll talk especially about Robert Heilbroner and his work, in particular his last book, *A Study of the Human Prospect*—take a very "sober" view of the future that awaits the people who will be living in the year 2000. Heilbroner bases his predictions on the conclusions of the Club of Rome and of his own economic and ecological analyses. He foresees, if not the disappearance of the industrial world, at least a diminution of its power under difficult economic and political conditions.

He also sees an orientation toward authoritarian regimes, which would be more capable of programming the survival measures necessary in any crisis. Paralleling this evolution will be the creation of monastic-like communities where intellectuals and others who want to go on thinking and creating can continue their efforts, as in the Middle Ages.

M.S. There are other scenarios for the future, in particular one that must irritate you since you are a world-renown specialist on the drug problem. I mean the scenario proposed by Arthur Koestler, who sees a euphoric world, similar to Huxley's, where man will use drugs and tranquilizers to adapt to the difficulties of his environment.

G.N. I believe that neither Koestler nor Huxley has given sufficient attention to interpreting neurophysiologists' studies of the past twenty years. I'm talking in particular about Koestler, who in his book *Janus* popularizes for the first time the work of Olds and other American scientists. Those men have exposed the ambivalence of the human brain, torn between the paleocortex—the old brain; the seat of instincts, of endocrine regulation, and everything that contributes to the immediate survival of man—and the neocortex, which consists of successive layers of neurons accumulated in the course of the two million years it has taken *Homo sapiens* to evolve into what he is today.

As Francois Jacob recently expressed it, there is a profound ambivalence between these two zones of the brain: between the new brain—which is able to make predictions, to think in symbols, to express—and the old brain, the seat of the instincts, of the urge to reproduce, of aggression. Konrad Lorenz did not integrate this model into his ethological works because it was not available to him.

The result of the brain's evolution is a certain equilibrium between those two parts of the encephalon. I believe that the "miracle" of the brain is that so complex an organ can engender what Claude Bernard called a cerebral homeostasis involving thousands of feedback and regulatory mechanisms that enable a "normal" individual to feel at ease in normal surroundings. Therein lies the miracle, which from the viewpoint of pathology ceases to function only in such extreme cases as schizophrenia or depression—cases, moreover, that do not require diagnosis by a psychiatrist, since any individual with a normal brain can recognize a madman who thinks he is Napoleon or a depressed person who wants to commit suicide.

Today, these two types of mental disorder are controlled and greatly alleviated by chemotherapy. Phenothiazines, for instance, make it possible to reduce the schizophrenic's increased production of dopamine in certain neurons. This treatment must be continued for a very long time in order to maintain cerebral homeostasis; but it allows schizophrenics to reintegrate themselves into society. As for depressives, they benefit from treatment by lithium and tricyclic compounds. A whole range of behavior is situated between these two poles. The study of human behavior reveals that on the whole humans, like animals, let their old brain predominate and hence seek pleasure. The great contribution of Olds, whom I just mentioned, has been to succeed in localizing in a very precise zone of the old brain the septal region of the hypothalamus: the pleasure center that is at the heart of animal behavior and probably of human behavior. His experiment consists of im-

Pleasure and Dependence

planting an electrode in the animal's brain and teaching it to press on a pedal that stimulates this zone with a very weak current. The animal will press on the pedal until he is completely exhausted even though it is possible for him, by pressing on another pedal, to obtain his favorite food. But he forgets the food pedal, preferring to give himself "pleasure." This experiment has also been carried out with monkeys, and similar observations on man have been made by Robert Heath.

M.S. It seems to me that José Delgado carried out a similar experiment. . . .

G.N. Delgado believes that human behavior is equally influenced by another center located in the old brain: the zone of aversion, the zone that induces behavior in reaction to fear. He decided not to remain in the United States, where the predominant philosophy holds that the only way to modify behavior is by recourse to positive reinforcement of the kind dear to Skinner; that is, a reinforcement that appeals to our wish for pleasure and satisfaction. This point of view is completely opposed to Delgado's "negative reinforcement." In my opinion, the Americans are making a serious mistake, because education must to some extent use both types of reinforcement, and, whenever the individual has the opportunity for it, the pleasure center becomes predominant. Olds says it beautifully: "Operant behavior is initiated, programmed, and terminated by rewards, either expected or obtained."

That statement contains the entire explanation of animal behavior. Since Olds's experiments, performed on animals and the rat in particular, biochemists have shown that the different phases of research on rewards were associated with biochemical modifications of the brain.

M.S. Are you speaking with an optimistic brain or a pessimistic one?

G.N. You speak of the brain, but first it must be defined. The structure of the human brain orients the individual toward seeking immediate pleasure at the expense of the pleasure he could have in the future. It is this search for immediate pleasure on the part of man that, in my opinion, raises doubts about the whole problem of futurology, especially in a democratic society where each individual can do largely what he likes. Thus, the margin of freedom enjoyed by everyone is employed to obtain personal satisfaction. The latter may be beneficial and contribute to improving the quality of life for the entire society; but that situation is far from being the general rule. It is tempting for futurologists to think that the neocortex is dominant in man and that, provided with a dream of a better world, he will be able to abandon his organic desire for immediate satisfaction. Nevertheless I believe it is difficult to change that fundamental orientation in man by chemical means. Koestler felt that man's sole chance of survival was to have recourse to a pharmacology that would make him forget his desire for immediate gratification and help him to find new orientation.

I believe Koestler made a mistake there, because he tried to interfere with a basic mechanism we do not completely understand. Twenty years ago, as a young pharmacologist, I myself was convinced that Koestler was right. I remember I discussed many of these questions with my friend Beuve-Méry. I used to say to him—this was before the time of antidepressants and tranquilizers: "We absolutely must develop a pill that will allow people to become wiser." Beuve-Méry laughed at me, and he was right.

M.S. Since you study the phenomenon of drugs with such acuity, I would like to ask you if you don't think fear of the future might explain why, since World War II, we have witnessed this fantastic popularization of what was before the war a kind of elitist perversion?

G.N. I believe that this widespread use of drugs—formerly called narcotics, now called euphoriants—is due above all to the fact that they act on the pleasure center of the brain and give the individual that immediate affective reward he is seeking.

M.S. We're back to the rat's pedal. . . .

G.N. Yes, but other factors are involved, such as the one you just mentioned. I believe that young people today are afraid of the future. That's why it's difficult to orient their new brains toward goals that would give them delayed satisfaction. They prefer immediate satisfaction. Americans call them the "now generation," an expression that I find perfectly accurate. I don't mean this pejoratively. I have a lot of compassion for their generation, and especially for those young people who, in spite of everything, think about the future. But they must realize what an ambiguous position they find themselves in.

M.S. But even without recourse to hard drugs, do we not now have available to us as a palliative to all our "troubles" an entire psychopharmacopia that the doctor can prescribe when requested? Not only that, but there have even been announcements of the discovery of drugs with very specific action; for example, those that could improve the capacity for analysis (predicted by some people for 1985), those eliminating anxiety. . . .

G.N. All of which makes it possible to influence millions of people, at the very least. This seems to me a very serious situation, espe-

Pleasure and Dependence   139

cially when I see, in France and the United States, the abuse of certain drugs like Valium. All the more so because we don't understand the mechanism of their action on the brain. These substances suppress anxiety. And yet it seems that anxiety is indispensable to all creative human efforts. If you had suppressed anxiety in Baudelaire, you wouldn't have had *Les Fleurs du mal*.

Recourse to psychotropic drugs is not viable in a normal person, even if that person is afflicted with the basic anxieties that are merely part of life. On the other hand, the benefit is evident for a schizophrenic who is being given phenothiazines. These drugs act on a precise point of impact, and result in chain reactions that will vary with the individual, unless there is a very basic lesion at the two poles we were talking about earlier. The normal person can, with these drugs, have his consciousness altered, with the extra risk of becoming a drug addict.

M.S. According to a list compiled by a number of American scientists [see page xvi], the discovery of a permanent mental stimulant is now predicted.

G.N. The use of physical stimulation to activate biochemical reactions in the brain is a fascinating field of research. Bob Heath, one of the first researchers to have isolated the pleasure center of the brain, has used implanted electrodes in the treatment of schizophrenia. The treatment of certain mental illnesses consists of programmed stimulation of certain zones of the cerebrum and cerebellum. I personally find this approach, which can be modulated, more interesting than the pharmacological approach. If I were to do research, I would orient it more in that direction than toward medication.

M.S. I continue my litany: the prolonging or shortening of memory; the suppression or development of the maternal instinct; the regulation of sexual responses; the improvement of sociability; reduction of the "need" for sleep; diminution or lengthening of perceived time. . . . A whole demonic pharmacopia geared to influence the mental process is promised us for the twenty-first century. . . .

G.N. You call that demonic. In my view, those pharmacologists are boy scouts who go out on maneuvers without knowing exactly what they're going to discover because they have no model. One of the main proponents of this technique is a professor of psychiatry and biochemistry at the University of California at San Diego medical school, Arnold Mendel, who is convinced that this pharmacology has merits. He formulates his theses in a very relaxed and seductive manner. But he has encountered difficulties. For example, he gave amphetamines to football players, and their performance definitely improved, but they became extremely dependent . . . Mendel is convinced that psychopharmacology can modify in depth the behavior of individuals. What we are most lacking, it seems to me, are drugs with a specific action. Everything you have mentioned up to this point represents feelings that correspond to a myriad of biochemical balances in the brain. To be able to change a feeling (pleasure, fear, etc.) implies that there must be precise receptors in the brain for each of those states or feelings. But I don't believe that such is the case. I believe that the substances that act on the brain have a very general impact. Moreover, those feelings correspond, I believe, to a certain equilibrium that I would not like to see disturbed. Psychotropic drugs certainly have a beneficial effect in a great number of pathological cases. But they have also led to abuses, to situations in which the individual, instead of taking on a certain number of situations which are after all common in a normal life, has tried to avoid them by recourse to a drug; and has ended up by suffering from it and becoming dependent on it. (I'm referring in particular to tranquilizers, Valium. . . .)

While the drug may have alleviated the person's anxiety it has also diminished his creativity, and in extreme case he resembles a robot.

M.S. Do you believe it possible, within the framework of a democratic society, to make provision against pharmacological manipulations of the psyche?

We can protect ourselves against a bad newspaper by not reading it, and against the radio or television by turning it off. We also have a margin of freedom from external aggression. But what will we be able to do against the mind-altering effects of drugs used in hospitals or clinics—drugs that we will be assured are harmless?

G.N. Government agencies do exist that take on the responsibility of controlling the use of medication. Psychotropic drugs should not be provided except by medical prescription and to alleviate precise clinical symptoms. They should not otherwise be used. Although we know that every drug has its side effects, we do not necessarily understand the mechanism of its action. It is extraordinary to think that a drug I am very familiar with, tetrahydrocannabinol (THC), which is derived from cannabis, not only stimulates the pleasure center of the brain, producing a tranquilizing euphoria, but at the same time, and with only a few billionths of a gram, alters the hormonal functions that control sexual activity. Clearly, the danger posed by such substances should be controlled by government agencies, by doctors, or by medical associations (to safeguard their profession). We must absolutely avoid situations giving rise to the individual's becoming his own doctor. Self-medication is bad medication—that's one of the foundations of medical education. Personally, I am deeply opposed to the proliferation of all these substances—substances whose mechanism of action, elimination, and side effects we know little about.

I believe there are two options. Either these drugs will become very popular and widely used (which I believe will do much damage to society), or else they will be regulated, either democratically via responsible bodies (pharmaceutical, medical) or forcibly, via an authoritarian government. It is inconceivable that this tendency toward dependence on drugs continue without ultimately endangering the future of the human species. I believe that twentiety-century man is intoxicated by all the technological conquests that have been made, in particular the conquest of space and the moon landing. And that intoxication has made him lose his perspective. I believe, like Teilhard de Chardin, that our future is inscribed in an evolutionary process begun several million years ago, and this must continue. But that prodigious evolution will only be thrown off course by little understood drugs, which after all are xenobiotics; that is, they (unlike our foods) do not enter into cellular metabolism, and are eliminated after biotransformation.

M.S. Do you see a danger in the fact that people take huge quantities of psychotropic drugs without feeling the least guilt? They often even feel safer. . . .

G.N. This is where the doctor's responsibility comes into the picture, and I believe that responsibility is very great. Today's doctors, comforted by the existence of miracle drugs—in particular, the antibiotics, the diuretics, and the tonicardiacs—thought they were also going to be able to develop what a psychiatrist friend of mine calls "a penicillin of the mind." But the brain is an extremely complex organ, with mechanisms and natural regulations it is dangerous to disturb. And then, too, there's the question of side effects. All psychotropic drugs pass through the placenta and can interfere with the development of a fetus by acting on the central nervous system of both the mother and the fetus. It has been shown that children born to mothers treated with phenothiazine have a high incidence of deformities. And, to go beyond these statistics in what is observable at their birth, one would have to be able to follow those children to the end of their adolescence; only then could one judge the development of their central nervous system—to see if it showed certain functional deficiencies: problems with learning, memory, etc. An alarm should be rung. Indeed, experiments have shown that the treatment of pregnant rodents with phenothiazines produces in their offspring types of abnormal behavior that have been regrouped under the rubric "teratogenesis of behavior."

M.S. That's the thing that astounds me—the repression, on the one hand, of the use of hard drugs, and the permissiveness, on the other, concerning the use of psychotropic drugs.

G.N. Such permissiveness is dangerous, and the paradoxical attitude toward substances of equal potential harm—the hard drugs, as you call them, and the psychotropic drugs—is troubling. In my studies of euphoria-producing cannabis derivatives, I was interested in discovering any effect they might have on the development of the cell—on its metabolism. I was one of the first to show that cannabinoids retard cell division by inhibiting the formation of DNA, RNA, and proteins. Following those discoveries, I studied the effects of other substances, the benzodiazepines (Valium) and the phenothiazines (Largactil), which are even more powerful. Those effects are all the more serious in that the psychotropic drugs accumulate in the organism and thus have a more long-term effect. Consider THC. The body takes a month to eliminate a single dose. Largactil and Valium, to take two more examples, accumulate in the organism, and especially in the brain, because they are liposolubles. It is in fact that liposolubility that is responsible for their remaining for so long in the organism—for that, and for the fact that after they have exercised their psychotropic power, they continue their action at the cellular level. That action is not felt by the individual, but it is nonetheless real.

Thus, the danger these drugs pose resides in their prolonged retention in the organism and their side effects on cellular metabolism. Very few psychotropic drugs are in fact eliminated rapidly. Those that are must be taken more often. Once again, then, one confronts a biological impasse, which goes far to explain the laboratories' tendency to manufacture psychotropic drugs of increased lasting power. One example is methadone, which acts over a more prolonged period than does morphine. But now that the side effects are known, pharmacologists are going to have to try to develop substances that act quickly, are eliminated quickly, and present no detriment to cellular function. Psychotropic drugs, acting on the hypothalamus, influence the secretion of sex hormones controlling spermatogenesis and especially ovogenesis, which is infinitely more sensitive to their effects.

Actually, the maturation of the ovum is part of a cycle triggered by very precise concentrations of sex hormones, whereas spermatogenesis is a virtually continuous process. The maturation of the ovum is indispensable to the production of a healthy ovum; and problems often arise when one modifies any biological process by the use of psychotropic drugs. Doing this research, I understood the wisdom of Freud, who said that the brain was in the service of the gonads. But the gonads are in the service of the brain, because the production of the hormones that control the gonads is programmed by the release of Guillemin's hypothalamic factors, and is likewise influenced by impulses from the cortex, but also profoundly altered by psychotropic drugs. The hypothalamic region is the point of attack for all the psychotropic drugs, because it is the region that conditions behavior. Finally, the use of psychotropic drugs by adolescents must be stopped, since it is during adolescence that the endocrine, affective, and intellectual systems are being structured.

M.S. You have a reputation as an uncompromising enemy of those drugs usually called "soft," like hashish, by comparison with hard drugs. On what do you base your thesis, which in a sense goes against the current?

G.N. Recent scientific studies have made it possible to establish that the chronic consumption of hashish causes alterations of function and structure in the lungs, the reproductive organs, and the brain. Inverterate smokers of hashish suffer from an obstruction of the bronchopulmonary tracts. Preliminary studies indicate that the smoking of marijuana is more carcinogenic than the smoking of tobacco, and more damaging to the pulmonary immunological system. It also causes disseminated lesions of alveoli associated with deposits of cholesterol, indicating a destruction of tissue. After four week's use of marijuana, a diminution of spermatogenesis occurs, as does an increase in abnormal forms of spermatozoids associated with a diminution in their motility. These bodily responses are linked to a disturbance on the hypothalamus-hypophysis axis—THC disturbs the formation of the gonadotrophins FSH and LH—and on the germinative epithelium. The result is intermittent diminutions of testosterone, a male hormone essential to the normal maturation of the spermatozoids.

At the level of the germinative epithelium of the testicle, the cannabinoids, whether they be psychoactive or not, inhibit the synthesis of nucleic acids and the proteins of the primary cells: the spermatogonia and the spermatocytes.

These two actions of cannabis can explain the formation of abnormal spermatozoids observed in smokers of hashish, as well as their oligospermia.

A single dose of THC (the equivalent of a hashish cigarette) given to a long-tailed monkey will bring about a drop in the gonadotrophins

FSH and LH and in prolactin, pituitary hormones that control the ovarian cycle. The injection of THC during the luteal phase disturbs the production of progesterone and causes anovulatory cycles.

Similar reactions have been observed in young women who smoked from three to seven marijuana cigarettes a week. They exhibited a high incidence of irregular or anovulatory menstrual cycles. They also exhibited diminutions of the rate of hypophyseal hormones—in particular, prolactin. Different amounts of cannabis (psychoactive or nonpsychoactive) have an abortive and embryolethal effect on the rat, the mouse, and the rabbit. Litters of newborn rats nursed for six days by rats fed THC will exhibit development and behavior abnormalities associated with alterations in the endocrine function.

Rhesus monkeys treated daily with THC exhibit a rate of abortion and prenatal mortality three or four times that of the control animals. Sensory defects and behavior indicating somatic alterations of the central nervous system are observed in the offspring of mothers treated with THC.

M.S. What is known of the effect of hashish on the brain and on behavior, and is it possible for a person to build up a tolerance for this drug?

G.N. Cannabis has a marked effect on the hypothalamic region, the principal seat of the brain's pleasure center, and this explains the tranquilizing euphoria of cannabis intoxication, which brings the endorphins into play (as do the opiates) but by a different mechanism.

In rhesus monkeys, the inhalation of cannabis smoke for three to six months causes permanent alterations in subcortical EEG curves, in the limbic structures, and in the sensorial thalamic nuclei. Abnormalities in the ultrastructure of these signs, in particular in the synapse and the nucleus, are also evidenced.

Brain cells, which do not divide, would seem to be particularly vulnerable to the inhibiting effects of THC on protein synthesis and the synthesis of nucleic acids. Three months' THC treatment of young rodents causes a permanent drop in their learning capacity and induces abnormal behavior. Rhesus monkeys raised in a colony and treated with THC exhibit, after three months, antisocial behavior characterized by immobility, after having earlier displayed behavior of persistent aggressivity. This phenomenon indicates the cumulative toxicity of THC on the central nervous system.

Tolerance of cannabis (resulting in the necessity of increasing the dose in order to obtain the desired effects) has been observed in all animal species and in man. The chronic user may smoke from five to

Pleasure and Dependence

ten hashish cigarettes a day. Withdrawal symptoms are not as marked as they are in the use of opiates because the body eliminates cannabis only very slowly. EEG alterations after weak doses of THC reveal a telescoping of the time spent in transition between waking and sleeping. The subjects then exhibit disturbance of the corporeal schema and periods of REM sleep that can become hallucinatory. Mental recall is also diminished by cannabis.

M.S. Do you believe that the use of hashish carries with it the risk of prompting the use of stronger drugs, thus favoring the dangerous transition to hard drugs?

G.N. I will reply by citing the results of a study made in the United States involving 6,000 high-school students: freshmen and seniors. It shows that the use of hashish precedes the use of other narcotics (euphoriant drugs like the amphetamines, cocaine, and the opiates) that activate the brain's pleasure center. It also shows that 26 percent of the users of hashish will use other drugs, whereas only 1 percent of those who have never smoked hashish use other drugs.

There is, then, significant statistical evidence linking the use of hashish and that of other narcotics.

Once the noxious effects of cannabis are known, is it really possible to call it a "soft" drug?

M.S. Do you see drug addiction as proof of the failure of our civilization?

G.N. I believe that in this context any failure would first of all be due to the ambiguity of the human brain's impulse to gain immediate satisfaction, very often by some excess—alcohol, tobacco, food. People forget that sobriety would permit them to enjoy better health. Any failing would also necessarily be tied to a fundamental deficiency in modern man's existence: the loss of a natural rhythm in life, and the impossibility of his getting regular exercise. Now, these two factors are very important in that they contribute to cerebral homeostasis. This concept is linked to the "turnover" of certain hormones, the endorphines, produced in the brain and associated with any feeling of satisfaction. It has now been proved that physical exercise favors that turnover and cerebral homeostasis. The latter, centered on the brain's pleasure center, vacillates between the poles of activity and of repose, much in the same way that well-regulated physical exercise brings a satisfaction similar to that sought by the users of drugs. Thus, I believe that people today should ascribe more importance to regular physical exercise. Loss of the natural rhythm of life is particularly observable in large cities, where people scramble madly to get out into the countryside during the brief periods of respite afforded by weekends and vacations.

All these extremely simple data have been somewhat forgotten in the quest for a utopian, or ideally developed, future. And yet it is with respect to these realities that I would formulate a view of the future: I would like to see the man of the twenty-first century physiologically and psychologically integrated into the natural environment. Only under such conditions will man's future be well and truly protected and furthered—much more than it would be by prosthetic devices or trips to the moon or Lord knows what else. . . .

M.S. As for that man of the future, do you see him as less aggressive, more congenial? Or will the population explosion and the modifications of life-style caused by the increased density of the urban fabric, by the increased scarcity of natural resources, etc., make him more aggressive than he is today?

G.N. I believe that the example of recent years and of large-scale urbanization provide the answer to your question. As soon as people are herded, as they so often are now, into industrial cities, they regress and become more aggressive than they would be if isolated on a farm, working daily in a natural environment.

M.S. But we have no choice. To make people more tolerant of one another, won't we have to resort to drugs? For lack of any other alternative?

G.N. It is possible to imagine an industrial society so dense that we might resort to such a solution. But, in my opinion, the use of drugs as a social control could be properly administered only in a totalitarian system. Now, the example China offers us proves that the individual's behavior can be altered *without* drugs, by the imposition of the totalitarian system alone. In that sense, the knout will do the job just as well.

M.S. It would be amusing to put together an anthology of the scientific predictions made fifty years ago. The limits of our knowledge may have changed, but we have always known, for example, that fossil fuel would disappear and that human beings would multiply. . . .

G.N. Were we wise, we would limit right now, as much as possible, the demographic explosion. Which is, for that matter, what the Club of Rome has been advocating for a long time. I don't know whether urban centers will continue to grow. I live in New York—a "megalopolis" supposedly in the vanguard of progress—and for twenty years the city has been declining drastically, and continues to sink. I believe that urbanization and the notion of grouping in big conglomerations are going to have to be rethought completely.

I'm rather inclined to think that we shall see, in the future, a return to a civilization that is more agrarian, more fragmented, with individual use of energy resources, as they tried to do it in China. I'm thinking of water and solar energy. Those types of energy are much more accessible to little groups than to big conglomerations.

M.S. In your opinion, is genetic engineering a promise of a Golden Age? Or does it augur an apocalypse? Along with the atomic bomb, genetic engineering poses one of the big questions of our time, and the scientific community remains very divided on the subject.

G.N. I tend to agree with my friend Chargaff, who feels it so far remains a game played by Boy Scouts, fooling around with things they actually know very little about. I doubt that these experiments will end in very profitable applications. Genetic manipulation has been employed for a long time. Quite a few animal and vegetable species are the result of genetic improvement, and progress can still be made in that realm. But, although genetic engineering has obtained remarkable results, it is still confined by biological limitations, constraints imposed by the natural world, the limits of the human brain, . . . and I believe we will find it difficult to overcome these factors.

M.S. Human activities are paradoxical. On the one hand, we are warned of the dangerous consequences of overpopulation. On the other hand, one of the most fruitful sciences of our time is gerontology. Is it possible to live 120 years? Is it desirable?

G.N. One hundred twenty years is apparently the maximum life expectancy programmed in human genes, but individual factors are of great importance. People who have a genetic tendency toward diabetes will never reach that age. But, given the problems of overpopulation, should we hope for such longevity in anyone at all? Should we impose ourselves on future generations, glad to see their ancestors disappear in order to have their turn at a place in the sun? I haven't the slightest idea. One hundred twenty years strikes me as a very long time. As it says in Ecclesiastes, "There is a time to live and a time to die."

M.S. Doesn't it seem to you, however, that increasing life expectancy should be an essential goal of science?

G.N. No, I believe that when an individual has lived to the age of sixty he has lived his life. Anything beyond that is a bonus. But I do have certain reservations: There are certain exceptional individuals who enrich humanity, and they deserve to live 120 years. It's too bad that Mozart died at age thirty-five.

M.S.   Are you in favor of euthanasia?

G.N.   In my opinion, euthanasia is a false problem created by modern science and by our wish to live on at any price. Artificial methods of prolonging life did not exist for the doctor of the past, who had both compassion and a sense of his *métier*. I believe that every doctor knows very well when the time has come for his patient to die. I shall always remember a colleague of mine who, without knowing it, had cancer of the pancreas with an icterus that was deep and inoperable. I went to see him to talk about a congress that was to take place several weeks later. Actually, I went to see him because his surgeon had told me: "Go and see him and say your farewells, because you'll never see him again." The next day, the surgeon gave him morphine by intravenous drip, and my friend went to sleep forever. I hope that, if the worst happened to me some day, a colleague would render me that service. Of course, I'm aware that there are limits and regulations, but it is above all a question of good sense and of heart.

This debate—so fashionable in France and the United States, among other countries—strikes me as artificial because it has been born of technical possibilities for prolonging life. Some people would have us believe that therapeutic relentlessness symbolizes the triumph of modern medicine.

The problem of euthanasia was much more serious in the time of Hitler, for example, who had publicly decided to eliminate a certain number of individuals he felt to be useless or corrupting, people he selected in a crazy way by jumbling together mental patients and entire ethnic groups, including Jews and Gypsies. Euthanasia was a pretext for genocide. But when people refer to euthanasia in the case of that girl in New Jersey who was kept artificially alive for whole years, they are overdramatizing.

The Hitler era weighs heavily on all reflections concerning euthanasia and eugenics. All scientific debate on the subject is warped by the extremely passionate stands that are taken. It is impossible to approach those debates and take a serene, scientific attitude toward them without immediately being subjected to a kind of emotional blackmail. . . . But, as men of science, do we have the right to sidestep problems of this importance because of our experience of history? I concede that safety measures are necessary. I believe they are inscribed in the code of medical ethics, which has recently been revised. Doctors have benefited, it seems to me, from a large margin for maneuvering in the matter of euthanasia. But the weight of the factors I mentioned is such that certain doctors prefer to take shelter under legislation or codes. But what is really at issue—even more than the behavior of the doctor, who is also a citizen—is the nature of the political state and the ethics it evolves. We must remember that genocide did not end with Hitler.

FLOYD BLOOM
The Brain Connection

AT 47, FLOYD BLOOM is the fair-haired boy of the American Academy of Sciences as well as the director of the Arthur Vining Davis Center of the Salk Institute, in La Jolla, California, one of the most important American laboratories devoted to studying the biochemistry of the brain.

There are predictions that he will win a Nobel Prize in the next few years for his work on endorphins—work that parallels and complements that of his friend and collaborator Roger Guillemin—and especially for his work on the next generation of drugs to control schizophrenia.

M.S. Does utopia, that dreamlike projection into the future, play a role in your life? Or, like many other scientists, do you fear or reject it?

F.B. If you had a healthy utopia, what then would we spend our time doing? In the utopian idealist perspective, if you didn't have to worry about health problems, you could then worry about creating new and interesting things for human beings to do. You could have a utopian world in which intellectual enterprises and the creation of new forms of intellectual endeavor might carry us off the earth into the stars, or to wherever it is that human beings have a potential to go.

We talk a lot about health allowing people to live longer but, in fact, as one of our studies just made clear to us, nobody actually lives any longer; it's just that more people get to live the same length of time. But why is it that people die at all? Why is it necessary? If there were a utopian world of health, important contributing minds could live for hundreds if not thousands of years, I think, to achieve the potential of human capability. Of course, it might be dangerous if the wrong person lived for a long time. If you really want to consider utopia in a serious way and factor out the skepticism, and the fact that human beings are still human beings, there will be people you wouldn't want to live ten minutes, much less ten thousand years. If you just take all that into consideration, I'd say utopia can't possibly be considered dangerous. Now some scientists like utopia because it gives them a kind of a kick, and some others are very suspicious about it and think that a scientific utopia, like a political utopia, can do a lot of harm. You have a magazine like *Omni*, which is a tremendous success, with a large circulation. It has a typical view of utopia, scientific utopia, almost science fiction, which modern readers seem to long for.

M.S. Isn't utopia particularly dangerous in the area of health? Raising false hopes of cures and longevity?

F.B. The dream of a utopia is probably as important as its achievement. If you decide in advance that utopia is dangerous and that you are not going to go for it, what are you going for? A short-term attempt to cure a particular cancer? Then a new toxin appears in the environment, then you'll have to make random searches for something to deal with that because you don't understand the fundamental basis of how the cells react with their environment and regulate their growth. That's the difference in the short-term versus the long-term perspectives on what you hope to achieve, and it seems to me, as a scientist, that the quest for utopia has to exist. You have to consider the ideal as a positive goal, because otherwise you are shooting at short-term, much more limited targets, in which the fundamental basis of understanding will always elude the seekers, unless they are extremely lucky.

M.S. How do you see man of the twenty-first century?

F.B. Granted that there is going to be a greater density of urban living and a greater rarity of natural resources, and if you assume that there is not going to be a chemical reconstitution of some of those natural resources, it seems to me that the natural tendency for the population in general will have to be toward more aggressiveness, because those who exist will have to defend their territory.

M.S. So we are going to see a century of race wars or explosions.

F.B. We have been through centuries of race wars and explosions, almost always based on the scarcity of one particular natural resource at a time, whether it was colonies to buy the goods that the kingdom produced or to go on a search for particular spices. There's always been a desire to go after whatever you thought those particular natural resources would be at the time. Gold has been one, and it varies from time to time. Oil is the one we all talk about right now. But it's been radio isotopes in the past, and it will be something else in the future. All of the habitable land is not under habitation at the moment, and scientific ways can be found of making uninhabitable land inhabitable. So it may be quite some time before we reach extreme density point, but there are going to be places like Los Angeles and New York City and Paris and London where most people would want to live. I would not feel particularly happy living in the middle of the desert somewhere. But it may come to the point where most people will not have that as a choice.

M.S. Does genetic engineering promise a Golden Age or an apocalypse to you?

F.B. We are very convinced that genetic engineering offers golden opportunities for the years ahead. We can, first of all, do therapeutic things with natural hormones that it is not now possible to do. The creation of adequate supplies of insulin and somatostatin through bacterial engineering is undoubtedly a major breakthrough. It's not quite worked out in detail, but it's clear it's going to be. We also think that it will be in terms of unlocking the secrets of how neurons communicate with each other, how cells in general respond to hormones that genetic engineering is going to be very important in determining the relationship between precursor hormones and the actual hormones that are synthesized and released, between the macro-molecules that act as receptors and how they execute the commands of the cell. With bacterial engineering it would be possible not only to understand how these things are created but how the genetic messages are controlled and regulated. And they will then provide a fundamental basis on which to look at pathologies in human beings when deficiencies of hormone production or response can be elucidated. My group thinks it's very important.

M.S. What's going on in your field now?

F.B. Most people are very impressed with the work done by the Howard Goodman/John Baxter Group, in San Francisco, on the discovery of human proinsulin, the cloning of the gene for the creation of somatostatin, and the work done by them to show the precursor-product relationship between the pro-opiocorten and the betaendorphine molecule. A lot of this work was done in a very short period of time with genetic engineering and would have taken a hundred times longer with standard techniques: protein peptide purification and solid phase synthesis. And that "old" approach would never have told you about the nature of the RNA or the DNA that cause the cells to do it in the first place. But, if you ask me what I think is possible in the near future, I think it will not only be possible to work out these kind of relationships with the eighteen other neuropeptides that have been discovered, but also to look at mechanisms to purify the receptors and determine how cells synthesize the receptors and how they regulate the synthesis of the receptors and the molecules that receptors combine with.

M.S. And what work are you doing in your laboratory?

F.B. Our lab is focusing on cell/cell communication: on the chemical transmitters and hormones that regulate the activity of cells in the brain. Our goal is to understand how the systems work together to regulate the way we respond to the internal and external environment, which is what we normally call behavior. We focus our research on neurotransmitter molecules and really restrict that to just a few of the dozens of classes of neurotransmitter molecules. For years we focused on a system run by the chemical norepinephrine so-called noradrenergic neurotransmission, and we understand a great deal about where these connections occur in the brain, the molecular mechanism by which they operate, and the behaviorial attributes that this particular synaptic arrangement can produce. We are now trying to understand when, in an awake brain, these cells fire to transmit these particular kinds of synaptic messages.

In the last four years we started work on the endorphins. There was a need to start at the beginning to work out where these substances are found in the brain, what kinds of circuits they are involved in, and what those circuits do, and then try to understand something about when those circuits are active in the behaving animal's brain. We take this information on the neurotransmitter systems and try to relate it to important classes of drug action, and that's how we get involved with the antischizophrenic drugs, the antimanic depressive drugs. We are studying a good bit now the effects of alcohol on the biology of the brain and which particular sets of synaptic messages are confused or changed by the presence of alcohol or something that alcohol does to the body.

M.S. The scientific community in America, and particularly in California, is involved with bioethics, especially euthanasia.

F.B. By euthanasia do you mean to relieve the strain of somebody who is ill with a pain caused by disease, or are you talking about cleansing the genetic basis of society? Because I think euthanasia is practiced today. It is just not talked about.

M.S. There is a movement in Europe as well as in America to find a legal basis for it, rules for it, and not to leave it to the discretion of the local physician.

F.B. Since the time when I worked in a hospital, saving people's lives with relatively primitive tools, we've come to understand a great deal more about ways to prolong life. The practice of life-prolonging procedures is such a major industry in itself that it is difficult not to use them in any given case. But, ordinarily, the same people who are seriously ill would have died at the beginning of the second day of their hospital stay instead of at the beginning of the second month of their hospital stay. All of which has caused a tremendous draining of hospital and financial resources and doesn't achieve much. There are only a few cases I know about where intensive application of life-prolonging procedures has actually achieved the restoration of function. Most cases you just get vegetables, noncontributory people.

Euthanasia sounds like we are developing a new procedure to kill off people who are in difficult straits. A more appropriate way to look at it is as a way to limit the application of some of the life-prolonging measures. It doesn't necessarily sound as if we are actively killing someone; we are just not actively prolonging a very desperate situation.

M.S. Modern medicine has brought about a certain number of problems that didn't arise before. Can we codify them? Could we make the giving of organs mandatory? I think it's worthwhile to think about it. That is what I mean by a "new morality."

F.B. To make it mandatory is perhaps a bit severe but certainly finding ways to motivate people to do so could be done in a free society quite easily, I think. It's much easier to donate a functioning organ than to be called upon every year to give money to charities that you don't wish to see supported. I think people could be motivated appropriately in a free society to do that.

M.S. Do you think psychotropic drugs and manipulation of the psyche pose a threat to free will?

F.B. Manipulation of the psyche strikes very close to questions that I often hear. People say, "You really understand how this drug works, but how could you keep a bad person from manipulating you through drugs?" My views are that it's only by understanding how things work and how drugs act that you have any freedom at all. Otherwise you are constantly open to manipulation by unknown mechanisms. It's only by pointing out how LSD or marijuana or alcohol or nicotine really manipulate the brain cells of your head that people will have the information to make a free choice. Without that they are captives of their habits and of their ingrained behavior, not really dealing intellectually with choice. They are dealing reflexively with visual responses. I feel that only information can provide the basis to make a real freedom of choice, liberty-type decision.

M.S. That's a very optimistic view; on smoking we have the information.

F.B. No, we've only got information on what it does to your lungs. We do not have information about what it does to your head.

M.S. And do you think that will change anything?

F.B. Well, if what it does to your head could be achieved without the something that does it to your lungs, then people could choose if they wished to manipulate their heads without the penalty of having to intoxicate their lungs. If a person wishes to manipulate his head with marijuana twice a day and achieves the same degree of pleasure with less risk than that derived from smoking cigarettes or drinking alcohol, I think that it's individual freedom of choice. If that person could be shown a nonchemical way of achieving the same mental state, that should also be a choice. It would take scientific wisdom and documentation to show them that these things are possible.

M.S. Can mental illness be helped through manipulations of the psyche, by psychotropic or electronic means?

F.B. I'm not sure we can talk seriously about mental illness if we consider all mental illnesses. Our effort is to concentrate on serious psychiatric illnesses, which would mean the major psychoses, schizophrenia, manic depressive disease. Those are also probably too general a set of terms to be talked about seriously as to appropriate therapeutic devices, because we don't really recognize how many varieties of manic depressive disease nor how many different varieties of schizophrenia there are. It's like talking about a cure for cancer. It all depends on which cancer you are talking about, when you catch it, and how far it has spread. And the same is true I think for mental illness. There may well be states of mental illness that are easily reversible spontaneously, and they can be reversed by a dietary manipulation in which a psychiatrist or a therapist happens to apply the right technique—whether it's self analysis or simply a calming period of protection from the environment. From that point on the person reverts to what we call a normal style of mental function. But there may be other cases in which the same therapist and the same person are not going to get a therapeutic response because the disease has progressed biologically to the point where it's not reversible on that level alone. At that point we are now relying upon chemical manipulation—psychotherapeutic drugs—to try to obtain the calming or the self-analytical effects. Those drugs have to undergo a great deal of improvement, not only because they now produce many side effects, but because they really don't reverse the disease process in a large enough proportion of the cases. Patients are calmed and are able to function to some degree in society, but they are not in any full sense restored to a normal human being. They are also dependent upon those drugs for the rest of their lives. In that case it's a half-way therapy or a palliative therapy.

By electronic means, I suppose you meant feedback kinds of manipulations. I don't really have any experience with that kind of treatment.

M.S. Can you imagine a world where, thanks to protheses and grafts, failing organs will be replaced like auto parts?

F.B. An organ bank could maintain a high level of functioning in people in a way similar to replacement of parts of an automobile. It's even theoretically possible to imagine that with genetic engineering you could reprogram the primordial cell that created the organ in the first place.

M.S. What ethical considerations would be involved?

F.B. I guess this relates to the previous question on euthanasia. When would you know that a person was contributing his organs of his own free will and was dying a natural death early enough to make the organs still usable to the bank, and when would it be a case where the doctor was terminating life prematurely in order to be sure that he got a healthy brain to transplant. It's a real problem.

M.S. How did you settle this issue in the state of California?

F.B. We just made arbitrary criteria, that brain death is the sign of death. Brain death is decided by the organized activity of an electoencephalogram. When that ceases to exist the patient is declared dead.

M.S. Then about the removal of organs?

F.B.   It has to be voluntary. They can be consigned by the survivors.

M.S.   You mean by the consent of the survivors?

F.B.   In some cases a person can write a will that will overcome the right of the survivors to decide his fate. But in many cases no such stipulation exists. In most cases no preexisting stipulation is there, so that the survivors are the ones who make the choice, rather than the person whose organs are actually being donated.

There are cases where law and medicine could work together. If people who really, truly wanted to make their organs available, there has to be a way found for them to protect their rights while they are alive but to also protect their wishes after they are gone.

M.S.   In the twenty-first century, will there be sizeable human habitats in space or under the sea?

F.B.   There might be important advantages to be obtained, strictly on the basis of the earth's population expanding beyond the capabilities of earth. I think there could be people in space, in or under the sea, acting as farmers to furnish supplies to maintain life, or to create manufactured goods that could only be manufactured in space where heavy machinery might be manipulated with much less effort, or where uranium could be extracted or power supplies could generate electricity for earth. That's one of the things a utopian society could achieve, to explore man's potential off the earth as well as on it.

M.S.   What possibilities do you see for immunotherapy?

F.B.   It's certainly a major approach to the treatment of cancers: the ability to find out why under certain conditions the body doesn't recognize an abnormal growth and allows it to continue. I guess I don't have an understanding of how immunotherapy could correct a failure of surgery.

M.S.   The invasiveness of surgery, for instance, is rather barbaric; drugs are also an intrusion of some foreign element into the body. Can one imagine an immunologic medicine where, just by playing with the natural defenses of the body, we can get rid of not only surgery but maybe drugs too?

F.B.   Well, if you think about the causes for surgery, you have surgery to remove tumors, surgery to correct defects of blood vessels, surgery to repair congenital defects. It's hard to imagine an immunotherapy that could deal with congenital defects.

Genetic engineering won't repair congenital defects, but it might allow you to understand how those congenital defects arise. You might be able to create a new atrium for somebody though genetic engineering but it would be harder for me to visualize how you would understand the failure of the atrial septum to close or the formation of a congenitally stenotic or some other valvular defect. I think it's possible to imagine an immunology that will paliate many kinds of things for which we now use chemical treatment or surgery, but probably not all.

M.S. Do you think increasing computer control of public health will lead to a police state?

F.B. I think there is a very real danger. If you insist, in a utopian society, that everyone has to appear every so often to have his systems checked with whatever the latest electronic device would be for scanning health status, that could be manipulated easily into a police-like state where only some people were told the truth about their health and others were allowed to languish. On the other hand it would certainly elevate the general level of health over a considerable period and that's probably worthwhile. There is always going to be the danger that some unscrupulous person or persons will manipulate the system. That's not anything new with modern society.

M.S. Maybe those people with the ability to achieve greater mental gyrations will only emerge when we take care of certain health problems that would have killed them off when they were in their cribs.

F.B. It's an important implication. You are moving to statistical input that is not strictly random, because we don't know what we mean by random in this sense. But we remove in a sense the statistical determination that is a very important part of evolution and substitute in its place something which is not the result of the random.

M.S. Do you think minicomputers could be implanted and used for biofeedback to control disease?

F.B. Certainly in the sense of abnormalities of the heart and abnormalities of blood pressure, finding a way to make a person aware of the abnormality at an early stage is certainly beneficial. You can control the state of premature atrial fibrillations by being aware of them and simply increasing the outflow of the vagus nerve to the heart. That can prevent a serious disruption of your heart rhythm. People who have high blood pressure can be made aware of their blood pressure rising, can be taught ways either to supplement the medications they are taking or to avoid the use of medications.

The most impressive results are Neal Miller's results with human beings. Neal Miller is a professor at Rockefeller University. He has been able by electronic means to help people with paralyzed spinal cords, for example, manipulate their blood pressure so that they can, just by electronic feedback means, make the blood vessels in their legs constrict. They can keep their blood pressure up even though they have no ability to control the blood pressure through ordinary means. Probably the same is true for epilepsy. There are some epileptics who anticipate the aura of their epilepsy, others who don't and therefore have no way of knowing when they are about to be subject to an epileptic convulsion. By electronic means that could be monitored so the person could have some signal of abnormal activity. A diabetic, I think, might have a subcutaneous electrode that would sense blood sugar and could know when he was either hyper- or hypoglycemic. That's one of the applications.

M.S. The problem would be how to go under the skin.

F.B. You could probably wear it like a contact lens in a nonvision part of your cornea and sample the sugar levels in the liquor of the eye, which is a very sensitive measure. Bestman is working on that . . .

M.S. Will the new biology provide answers to agricultural, industrial, and technological problems as well as therapeutic ones?

F.B. I would say yes on food: I think that it is possible for the new biology to come up with ways to synthesize food or materials of that kind. I think it's possible to have the new biology come up with new forms of energy. After all, all of the petrochemical fuels we have now are biological in origin. It just took nature thousands of years to make them in a form where we could burn them. I think we can do the same thing. The body has enough electricity in it to light up several bulbs, if you only knew a way to harness that, having a microbacteria that could make an electrogenic pump molecule for example. You could take the whole sea and make a tremendous generator system out of it. I think that would be possible. Industrial needs, well, we talked about the pharmaceutical industry. The new biology can certainly contribute to that industry.

What I think the new biology really represents is the application to biology of the rules of physics. As we understand more and more about the atomic and molecular nature of the cell and the approaches that we have made to come to it, we now understand how physics applies to the cell system. We then recognize from that set of knowledge how to apply it back into the universe, from which the industrial applications of physics are the major ones.

M.S. How do you see the role of the doctor and of medical power in the future?

F.B. Certainly there is a great deal of covert medical power now, there is a great deal of idealistic medical power now, and the doctor is given attributes that he doesn't justly deserve. But I think the medical doctor has always been important because he is the person to whom you entrust your life. As the general level of health improves, it should be less and less necessary to have doctors to deal with crises and more and more helpful to have doctors to help you maintain your health. I'd see us going more toward a Chinese style where the doctor is expected to keep you healthy. When you go to him because you're sick, it's a failure of his ability to keep you healthy.

M.S. Is it possible to have noncoercive preventive medicine in a free society?

F.B. It's hard to conceive of preventive medicine that is not coercive, because human beings as a rule don't like to be told what to do. Even if you explain to them in great detail why something is good for them or something is bad for them, the nature of being a human being is the ability to make a decision.

An interesting example is Salk's vaccine. People choose not to take the vaccine; you have the Amish community where they get polio because they choose not to take the vaccine. The preventive cure is available, totally acceptable, and totally successful, and people still have the choice not to do it. Now if you could limit them, if you could let them live in a plastic bag and they could choose to catch polio if they wanted but not serve as a source of polio virus to infect the community, then I think it's acceptable to ignore preventive medical cures. But when it reflects on the propagation of a disease to someone else, then I think you are entitled to be a bit coercive, as we were in the case of the Amish people.

M.S. After all, the nucleus of the problem is to have a democratic society.

F.B. In a democratic society you have responsibilities, not just rights. I think it's your right to make a choice for yourself, but it's your responsibility to make a choice for yourself that doesn't impose something on somebody else's rights.

HENRI LABORIT
A Futurology of Happiness

HENRI LABORIT is the most brilliant "maverick" on the French intellectual scene. Born in Hanoi on November 21, 1914, Laborit first abandoned a military career in "la Royale" (the French navy), then a surgical career (hospital surgeon in 1948), for basic research. In 1951 he introduced the first tranquilizer, chlorpromazine, and artificial hibernation, and in 1957 was awarded the Albert Lasker Prize by the American Public Health Association. (It is called the "little Nobel," but is often worth the big one.) Henri Laborit has taken a particular interest in organic reactions to aggression. In much the same spirit, he has tried to extend the structural laws of general biology to the social sciences, building a bridge between physics and language. Hence a certain number of general theoretical works that, along with his specialized books, deal with human behavior in social situations. Since 1958 he has headed up the Eutonology* Laboratory at Boucicaut Hospital, which is run by a non-

*A word coined by the philosopher Canguilhem meaning the science of the restoration of equilibrium after an aggression of any kind.

profit organization. This laboratory, financed by royalties from patents taken out by the organization, functions without any aid from the state and provides employment for eighteen staff members.

Finally, Henri Laborit has published some twenty scientific works and findings from more than 700 original experiments. He is the founder and publisher of the international journal *Agressologie*.

But this astounding Who's Who rundown still doesn't adequately convey the richly energetic, ambiguous, generous, prolix personality of this man who is not only an internationally known scientist but an accomplished writer, a pamphleteer, and even—recently—an avant-garde filmmaker.

As I said, he is a maverick.

M.S. As a theoretician, experimenter, and not only a scientific but also a social innovator, and as a writer and poet—there are some very lyrical pages in *In Praise of Flight*—and now as a filmmaker, don't tell me that utopianism—or at any rate its younger sister, futurology—has not played a role in your life and your intellectual journey.

H.L. You can't practice futurology in a laboratory. Practicing futurology means forecasting a scenario for the future and then trying to achieve it. You realize that as soon as you start experimenting new facts emerge. The hypothesis—the scenario—is indispensable at the outset, but in the meantime new facts emerge that constantly change the perspective. You can't practice futurology; you practice *approximate* futurology, step by step. Fortunately, that's what makes it possible to discover things. Sometimes—but this is the exception—you can imagine something and then find it to be true, but that's very rare. Utopianism, on the other hand, is the hypothesis upon which all work is founded. Utopianism is practiced all day in a laboratory. Utopianism consists of starting from accumulated facts, imagining a new structure, and deploying the imaginary, the driving force of all creativity. The utopia itself is not realized, but it acts as the basic incentive behind all research.

M.S. In other words, it's skepticism that is intellectually sterilizing? Strictly as regards health, what do you think of our prospects? (I no longer dare speak in terms of "reasonable futurology.") There have been two great basic discoveries in the history of medicine: Jenner's vaccination; and antibiotics. What do you see, today, that might lead to great therapeutic innovations tomorrow?

H.L. Therapeutics in its present form is something I don't believe in; it serves only emergency medicine, and that isn't health. Let's take

A Futurology of Happiness

the example of antibiotics. If I had pneumonia, I would certainly want to be given penicillin. But antibiotics will not eliminate the causes of my pneumonia. Why did I get pneumonia? First, because I found myself in a position where I was frustrated—my ability to act was inhibited. I secreted glucocorticoids and norepinephrine. And we know that glucocorticoids destroy the entire immunological system. Thus, we should not treat a defective organ without first understanding the organism. Moreover, we should be able to "convoke" the entire familial, social, and professional environment of the patient, and evaluate his relations with it. But even this approach would be insufficient, in that each individual reacts to his environment, at any given moment, by bringing to bear all his prior knowledge, whatever has been stored in his brain—all his value judgments, all his prejudices, and everything he has been told was good, bad, beautiful, or ugly. In his nervous system are coded routes that run at cross purposes, as a course of action is embarked upon, with other routes that are coded differently. Thus, one constantly "negotiates." This must never be overlooked. The phenomenologists say one must take into account people's experiences; otherwise science is a joke. And they are absolutely right. And yet they are wrong, too, because it's impossible really to penetrate people's experiences. The individual will be able to reveal nothing but his conscious mind via a crude system of communication, language.

Thus, the treatment of illness must be social, or there will be no treatment at all. Today, medicine is simply engaged in a job of maintenance.

M.S. And, if I take your meaning, up to the present the triumphs of medicine have involved reinforcing only the crudest lines of defense, such as microbial aggressions, etc.

But we're no longer at that point. Let's talk about the second line of defense, since people *have* to be treated, even crudely, in the absence of the more complete, social approach of which you speak. I want to talk about the cellular abnormalities associated with cancer, and with hormonal and nervous disorders. You are a specialist in nervous disorders, and there have been breakthroughs. But we still don't know how our increasingly intimate knowledge of those disorders and their treatment—to which you have contributed much—are going to help us fight, even in the most limited sense, such things as mental illness.

H.L. Right now I'm looking for a molecule that would act on an enzyme to prevent inhibition of action. When I find it, I will block all that biological, hormonal, and nervous "bustle" triggered by that inhibition. But how will that success change the individual's relation to his environment? I will have transformed the soft into the hard, the vague into the definite, the passive into the aggressive. But the individual's new assurance and energy will open the way to even more conflict than he was forced to confront when he was more passive and resigned. Which will no doubt increase the number of suicides. When a man has his back against the wall, the only effective medication is a drug for the part of the brain that conceptualizes the unreal, the imaginary.

M.S. Such drugs exist, from marijuana and hashish to morphine. But, with the exception of morphine, they aren't used in therapeutics.

H.L. Yes they are. Therapeutics, in that kind of situation, resides in psychotropic drugs, including hashish, heroin, etc. They have the major disadvantage of causing phenomena of dependence, or addiction, and the terrible pathology that flows from that, but otherwise. . . . We need drugs that encourage the imaginative process. The only way out, when you are caught in a Manichean system—a system that gives you no other choice than Scylla or Charybdis—is to find another solution. But people can't find it. They're caught in a trap.

M.S. Therapeutics via escape.

H.L. It cannot avoid being an escape. But above all it's come down to a matter of imagining a kind of behavior that will release the individual from those two impasses. Now, discovery supposes an imaginative process. Imagination comes into play in artistic activities but also in the necessary remodeling of everyday life with respect to others and the environment. It's very difficult, but that remodeling is the only way to act therapeutically.

M.S. In your view, wouldn't the cure of cancer be a basic discovery?

H.L. Yes, but only if there is a cure of the pathogenic process. Now the pathogenic process exists and spreads because an increasing number of individuals are frustrated. I'm surprised at how hard it is to get this message across to my contemporaries.

In this complex system that is life, there are no causes and effects. There isn't really a cause of cancer, just as there is no cure for cancer. The same microbial strain injected in different individuals will provoke a boil in one, septicemia in another, and nothing at all in the third individual. That's why what used to be called "the terrain" is basic. Claude Bernard used to say: "The microbe is nothing, the terrain is everything." And he was right. At the time, however, it was believed that this "terrain" was genetic—that there were people who were born to be sick and others who were born to be healthy. But that's not true.

A Futurology of Happiness 165

There are people who, since their childhood and owing to the history imprinted in their nervous system, are either capable of regulating their environmental situation physically and psychologically, or incapable of regulating it. And as long as we refuse to approach the problem on that level, there will be no cures of cancer or of anything else.

M.S. You're a Jansenist of health. One either has the "grace" to be in good health, or one doesn't have it.

H.L. All irony aside, it has nothing to do with grace or good works. It's a matter of environment and hence of society. Why are the most disadvantaged members of society the sickest? It's not because of physical or mental predispositions but simply because, in general, they are the people who have the least chance of satisfying their needs—for giving themselves pleasure—and who are the most inhibited in their actions. It's no accident that a big executive who is aging a little and feels that his authority is beginning to be challenged, has a heart attack. Not even a broken leg happens by chance. . . .

M.S. You are refining the notion of "stress" as popularized by Hans Selye.

H.L. At the time when Hans Selye talked about "stress," the word seemed to defy translation. That's why I never used it. In my first book, in 1952, I used the expression "organic reactions to aggression and shock," because I couldn't find a French-language equivalent for "stress." Today people use the word "stress" in the sense of aggression, which is not accurate, or at any rate is inadequate. One cannot conceive of a living structure that is not in constant conflict with an environment that is less complex than itself. Merely by virtue of the fact that it keeps itself alive, it is "under stress." In a way, stress is life. For me, life is stress and memory; that is, the simultaneous maintenance of a complex structure and imprinting via protein synthesis of past experiences. It's a somewhat personal definition. A living system is a system that registers experiences in such a way as not to repeat those dangerous to it, and to repeat those that are favorable. Even an amoeba has a memory and is a complex structure that, so long as it lives, has its autonomy, its structural particularities with respect to an environment much less structured than itself. The notion that there are no causes and effects in a complex system but rather multiple causes that have multiple effects through interaction is a fact that hasn't yet struck home. The cyberneticians keep repeating it, and yet people continue to believe that the identification of a causal agent will lead to the solution of a problem. Here's an example to illustrate what I am saying. Imagine that animals delivered aseptically from their mother by

Caesarian section are immediately placed in an aseptic environment and fed with aseptic food. When they reach adulthood, those animals, having never encountered a microbe, are put into a normal environment in which other animals are living perfectly well. What happens? In a few hours or days, the first animals die. What is the cause? The microbe? Or the absence of microbes?

M.S. You are no doubt aware of the futurist, utopian tables drawn up by highly respected scientific institutions—the Royal Society of Medicine and the Stanford Research Institute—and by scientists of international reputation whom you know well, like Bender, Blum, Ebright, Evans, and Gabor. In the field that particularly interests you—that of molecules with psychotropic action—scientists have predicted the control of aggressiveness, of memory; improvement of the capacity for analysis, for learning; the stimulation of intelligence, of sociability, of sexual appetite, and even such curious things as "a deeper awareness of the beautiful" (Gabor), "the development or suppression of maternal behavior" (Evans and Kline), etc. And all that before the year 2000; in other words, tomorrow. Is this merely science fiction?

H.L. No, not at all. I think it's all very amusing and, for the majority of those predictions, within the realm of the possible. But I don't believe in the social advantages of pharmacology. I don't believe that man, in making drugs, will transform himself in any permanent way. Only by knowing himself, and not by using a "crutch molecule," will he succeed in transforming himself for the better.

M.S. In a way, when you say this it's as if one of the fathers of modern psychotropic drugs is disowning his children.

H.L. No, I'm not disowning them. Let's say that I'm "relativizing" them, or putting them in their place. There are patients who, given the present state of our knowledge, need those drugs and derive great benefit from them. Obviously, the case of a maniac with criminal obsessions needs pharmacological treatment. No logical discourse can touch him or bring him back to a normal level of awareness and self-control. Once "cured" by medication, such a person returns to the same familial and occupational environment that he fled in his madness. We should not then be surprised to learn that he has again killed. He will be put back in a psychiatric milieu and again treated chemotherapeutically. The important thing is to know how to administer doses and otherwise use any medication knowledgeably as part of an overall strategy. If the person in question had not, from childhood on, lived in an environment that prevented him from expressing himself, he would not have acted as he did, and would not need medication. Actually, the medication treats behavior that is the result of experiences going back to his birth.

M.S. If, tomorrow, man is better able to untangle that skein—if he is better able to understand himself—will he really be better off? In other words, do you see twenty-first century man as less aggressive and more congenial? Or will the population explosion, the changes in life-style, the increased density of the urban fabric, and the increased scarcity of natural resources make people still more aggressive than they are today?

H.L. I come back to cybernetics. We must remember that man multiplies because he is able to do so. But that phenomenon will come to a halt in and of itself when conditions block it. When you put animals in a cage, and the cage's population reaches a certain density, they no longer reproduce. Man is not isolated in nature. He transforms his environment, which in turn transforms him. Thus, insofar as the man of the twenty-first century is concerned, I don't know what he will be or will not be. He will be what a complex of infinitely numerous factors make him. As regards human behavior, futurology is extremely inexact. I would say only that humanity is an animal species like any other, and that as such it will perhaps disappear. If it survives, it will have found a new way to live and will have reversed its suicidal march, its aggressive behavior toward itself and the biosphere. Basic aggressiveness is competitive aggressiveness. Which means that from childhood on you don't encounter other people but rather competitors among individuals, groups, states, and blocs.

M.S. Among the great scientific breakthroughs of our time, I believe that genetic engineering is in the first rank. Today, we are witnessing the emergence of new prophets who declare that genetic manipulation—in the animal kingdom as well as in the vegetable kingdom—promises a Golden Age. Others foresee—as they did in the discovery of the atom—an apocalypse.

H.L. Certain genetic illnesses—for example, those due to an enzymatic deficiency—constitute one of the many areas to which genetic engineering can be applied. Certain individuals will be able to have a normal life thanks to that molecular transplanting. I believe that genetic engineering is important, and that it will obtain rapid results. This is far however from saying that it is going to transform the human species and health. I might add that these manipulations are not without their risks. Selection pushed too far will mean the disappearance of a certain number of species, both animal and vegetable. And once the genetic pool is considerably confined, evolution is no longer possible. This is a general law of life that man should not violate.

M.S. Is it possible to live 120 years? Is it in fact desirable?

H.L. It has been said that at the age of thirty-five, a person can no longer create. That's entirely false. But the idea is so well anchored in people's minds that in fact creativity is blocked very early. What is at issue is not age but the assumptions, the conditioning, the value judgments that man is subject to all his life. "To add life to years and not years to life"—that's what's important. I believe it is theoretically possible to live 120 years. Each species has its life span programmed, so it seems. Perhaps we have a gene that determines the moment of the end of life. That would be a case of genetic and hereditary transmission. The immortality of the isolated cell would no longer be conceded; we know that a stem divides a certain number of times, about fifty or sixty times, and then perishes. If you introduce an antioxidant into the environment, you see that although the aging is genetically programmed there are no longer sixty divisions but double that number: 120. Thus, the life of a clone is considerably prolonged. I think that's what we're going to arrive at—relatively soon, for that matter—but it's interesting to understand the why of the oxidation. Corson's work is enlightening in this respect. A frustrated animal doubles his consumption of oxygen and his production of CO_2. He releases norepinephrine and glucocorticoids, and consumes more oxygen. Thus, the more we are frustrated, the more we age. Considerable releases of norepinephrine and glucocorticoids prevent REM sleep, which is necessary for good recuperation. The hope for life has now reached a limit. The advances made in medicine—the elimination of great epidemics, of tuberculosis, and the conquest of infectious diseases—is nullified by a way of life and an environment that are more and more pathogenic.

M.S. Can euthanasia be integrated into a new morality?

H.L. I believe that euthanasia is the kind of question that allows Western people "of good conscience" to avoid real problems and to divert public opinion from them. Words, value judgments—none of them to be taken seriously. All human life in the West is based on a language of evasion . . . War and hunger are basic problems that our West avoids. Our attention is constantly diverted from these real problems toward questions of a pseudometaphysical nature, which are secondary. Euthanasia is already being practiced on a global scale on the peoples of the underdeveloped countries. Wars, genocide, and the exploitation of man by man are more basic problems than euthanasia.

M.S. Do you believe that words are meaningless? Aren't you

yourself making a very explicit value judgment on the society we belong to? I mean, liberal parliamentary Western society. For all the reservations one may have about that society—with everything in it that is perverted by money, certain elites' appetite for power, etc.—is it not still more tolerable than the totalitarian societies, more respectful of the individual and hence of his life and suffering?

H.L.   Not having any money, I have nothing to do with your liberal society. I'd say that the totalitarian society is nothing more than another approach to establishing a dominant power. People in the West have discovered a different way of establishing this dominance, and that's through conformity to a certain ideology. This form of totalitarianism is worth about as much as the other. The essence of our fine, liberal, nontotalitarian society is based on the notion of property. There is no zone in the brain where the notion of property is inscribed. It's something learned; it's tied in with a system dating from the Neolithic period in accordance with which all our laws are founded. Those laws were conceived only in order to defend it. I don't see any difference between one society and another. Either private property reigns—and if you don't agree with the laws you're a pervert or a delinquent, and they put you either into prison or into a psychiatric hospital—or else the property is not private but belongs to a state managed by bureaucrats instead of the bourgeoisie, with the opposition being similarly sent to a psychiatric hospital. The main distinction has to do with learning different rules by which to maneuver. Until we understand man's cerebral functioning, each person will continue to express his aggressiveness and to defend what he believes to be the truth, because that truth is the one that most gratifies him.

M.S.   That truth is always subjective. It seems that even the person least provided for has a better chance of flourishing, and more freedom, in a nontotalitarian society than in a totalitarian one.

H.L.   People believe that happiness has something to do with the possession of objects. The individuals at the top of the ladder congratulate themselves and take their gratification in other forms: power and fame. The notion of freedom necessarily ends in intolerance. Intolerance is the daughter of freedom. One person enjoys the certainty of possessing a truth chosen freely, while the other person has chosen error—just as freely. So he has to be done away with. The top executive and the bum are both free to sleep under a bridge, but it's only the bum who takes advantage of that freedom.

M.S.   Do you not even believe in what appears to be a consensus derived from a vote, for example? From the system of political determination we call democratic, in spite of all the cheating and all the historical and sociological unwieldiness?

H.L. How are you going to vote? How can you believe that you are voting "freely"? You have an image of yourself that you try to project onto your social milieu, and it is reflected back to you by that milieu, favorably or unfavorably. If it's unfavorably, you will of course want to change the social structure so that your merits are better appreciated, and at that moment you will vote for the left. Or else you are rewarded with the Legion of Honor and the Cross of Merit, along with a promotion, and in that case why change the social structure? It must be preserved, since it recognizes your merits, so you vote for the right. Such, it seems to me, is the play of democracy, and therein lies the illusion of democratic freedom.

M.S. What I don't understand is that what you describe with such ferocious and yet good humor applies much more to nonliberal societies than to liberal societies, which accept challenge and questioning. They tolerate being destroyed and rebuilt.

H.L. I don't think that people will destroy liberal society; it will destroy itself. All societies destroy themselves because they want to persist in their errors. What is wrong is to want to persist in a feeling of superiority, to want to preserve the hierarchical system of dominance. As for reconstruction, it is difficult when one has not yet understood why the system holds together at all.

M.S. A democracy is the only society that is constantly asked to legitimate itself. You strike me as very unjust in the way you look at men as if you were a coolly removed entomologist. It's enough to make one despair. Is there no other solution than suicide?

H.L. That will be the only solution when there are as many cops and bureaucrats as there are "free" people, using your definition.

M.S. You are one of the pioneers of modern psychotropic drugs. Does man have reason to fear that his free will, his freedom, will be seriously threatened by the new psychotropic drugs? Is it possible, within the framework of a democratic society, to make provision against excesses of that nature—the manipulation of the psyche? And if so, how?

H.L. People have nothing more to fear because they are manipulated every day, constantly, without psychotropic drugs. Day in and day out man is molded into what is wanted. Even from the viewpoint of the family structure, man is manipulated from early childhood. He is inculcated with notions of good and bad, what he should do to be rewarded, what he should not do to avoid being punished—notions that, as a matter of fact, vary from one period and region to another period and region. Throughout history, the slave has not been recognized as a human being; and one wonders whether, even today, in certain countries to be black is to be considered human. The Indians in America were killed as if they were not human. One doesn't need psychotropic drugs to alienate or to be alienated: The mass media suffice, since they are in the hands of the powers that be.

M.S. But that's not what I was talking about. I mean to say that the products you helped discover are being put to increasingly abusive uses for repressive ends, both by the left and the right. The phenomenon is constantly spreading. You yourself are one of these who developed a "sexual disinhibitor," which is now on the market. What is the advantage for the individual, for society, of the discovery of such products?

H.L. My team and I are accepted because we do in fact discover products. That productivity of our laboratory is of lively interest to the world of commerce, which hopes to sell what we make. I have let myself be greatly exploited, but I suppose that's a characteristic of the merchant society in which we live. In that context, I really have no choice. To come back to your question, I don't see much danger in psychotropic drugs. I believe that once again, as in the case of euthanasia, psychotropic-drug abuses are a false problem, emphasized in order to deflect attention from the real problems; in this instance, the problem of human behavior. The powers that be have always found, even before psychotropic drugs, efficient ways to establish and maintain themselves.

M.S. Can mental illnesses, at least, be treated through manipulations of the psyche, whether by psychotropic drugs or by electronic means?

H.L. A person who has what is called a mental illness—a neurosis or a psychosis—no longer carries out his productive function with respect to a certain social environment, which has determined what is to be considered suitable behavior. To make a new society, one must have new people. One cannot exist without the other. No change can be hoped for until people become aware of their alienation from the power structure, of whatever nature, and until a psychological revolution has taken place. People must understand how their brains function and how their exploitation is effected in an alienated and alienating society. People should be given the means to be happy without psychotropic drugs.

M.S. Couldn't psychotropic drugs be rigorously controlled in order to prevent abuses and the dependence they induce? Here we touch upon the problem of the power, but also the duty, of the doctor.

H.L. Probably, but that's of no interest. Whether psychotropic drugs are regulated or not, with them people will reach the "best of all possible worlds" and have no more worries. They'll be happy. In India, people put up with being untouchable. They are tranquil, because they are sure of returning in another life, having moved up one caste. The churches and the powers that be have no need of psychotropic drugs to perpetuate their domination. Even if we were to discount the "visible" seats of power, a whole range of very useful assumptions assert themselves. The father of a family responsible for the education of his son will decide whether he goes to a lay school or a parochial school, etc. The *invisible* structure that substantiates the ruling authority perpetuates itself indefinitely. . . . Even alcohol, a classic psychotropic drug, plays a role.

M.S. That brings us naturally to another aspect of consciousness—sexuality. What kind of sexuality will we have tomorrow in a world where, from the contraceptive vaccine to the test-tube baby, conception and sexuality will be totally dissociated?

H.L. It situates itself easily, without difficulty. There is an impulse, and on that impulse is grafted an automatic sexual response. For centuries we have been taught the following scenario: The woman "possesses" the man, the man "possesses" the woman who "possesses" him, as if love were tied to possession. Thus, a person is prohibited from using, in his sexual life, the most human element of his being, the associative brain, which allows him to create and imagine. Man and his sexuality are encased in an iron collar of value judgments, assumptions, prohibitions, habits. . . .

M.S. Then, in your estimation, what kind of sexuality will we have tomorrow?

H.L. Everything that is possible without hurting the other person, and with the other person's consent . . . but this "other" person must, likewise, have left behind his or her own preconceptions.

M.S. To rediscover conviviality in love, in a way. . . . Do you see the beginnings of a new sexual morality?

H.L. Or, rather, the beginnings of a new kind of behavior. Man should be able to express his sexuality outside the taboos and clichés society imposes. Sexuality should be dissociated not so much from conception as from the sense of possessiveness that gives rise to failures and disappointments. I'm thinking, for example, of communes.

They usually fail because the reflexes, the assumptions, of "sexual property" take precedence over the ideology on which the communes were founded. Sexuality should be a creative impulse admitting the intervention of the brain. In a way, we should invent a new dialectic of love that would be redefined for each couple. Everyone should imagine his or her sexual behavior so that all routine, all repetition, is excluded. Sexuality has meaning only if it involves the constant discovery of the other, and even more so, of oneself. Primary narcissism shows through in the sexual act; then it becomes a desperate search for the self, beginning at the moment when one becomes aware of solitude, of birth, and of death.

M.S. Data processing has entered our lives, and tomorrow it will play an even greater role, particularly in health care. It should soon be possible, for example, to devise our genetic, or biological, identity cards. . . .

H.L. What's worrying you? Data processing is an instrument. You could ask yourself the same question about railroad timetables.

M.S. No, I think it may be the prelude to a still more policelike society. Not only the Inspector Maigrets of the future—which is not so serious—but also the Gestapos of the future will have very powerful and refined means of identifying individuals with perfect false papers.

H.L. Your remarks permit the supposition that the police system would go unchallenged. You seem to grant that the hierarchical power structures, of whatever kind, will perpetuate themselves. On that subject, I am neither an optimist nor a pessimist. Data processing in the hands of a totalitarian and coercive power can be frightening. But what is at issue is not data processing, which is an instrument and nothing more, but power—all the forms of power. Up to the present, technological evolution has made it possible to maintain and reinforce the power elites. But today I begin to believe that this technological evolution might fall into the hands of those who support the progressive and slow destruction of all power. Success will depend on the capacity to break the present monopoly of information networks. The latter will become so complex and diversified that those in power will soon be outflanked and will lose total control of the system. Actually, the censorship of thousands of parallel channels of an information network becomes an insoluble problem; the proliferation of pirate radio stations illustrates the phenomenon very well. Haven't the Italian authorities, for example, thrown up their hands?

M.S. Man will be able more and more to control his body through electronic means. Diabetics and heart patients will probably wear microprocessors placed in a watch-case to control their glycemia and their heart rate. What do you think of such prospects?

H.L. Fine, I'm all for it, but it's merely technical progress. Gadgets will not change humankind. They will improve the fate of the diabetic and the heart patient, but we must think on the level of the species and no longer of the individual. The overall mass of men on the planet, who are killing one another everywhere and who are dying of hunger here and there, poses far more serious problems than those you have raised. The watch-cases you mention will contribute to the individual health of the patient but, more to the point, they will make the manufacturer's fortune. This technical evolution will be effected at the expense of the consumer, who will be dependent on these devices. It seems to me that such gadgets constitute one more instrument of constraint, of coercion, of alienation.

M.S. People expect so many miracles from the "new biology" that some of them believe it will provide answers not only to our therapeutic problems but also to the nutritional, energy, industrial, and other needs of tomorrow's world. Do you believe it will?

H.L. I'm distrustful of the "new biology" because I'm aware that the manipulation of living matter can induce unexpected and sometimes grave disorders. I'm thinking of Egypt, whose industrialization took the form of the Aswan Dam. The decision was made, and the dam was built. But the other side of that coin was a large-scale ecological disaster associated with an unprecedented diffusion of bilharziasis (schistosomiasis). Since then I've been very prudent. Messing about with living matter, natural cycles, can lead to such a boomerang effect. Yet, biological research can contribute considerably to the solution of nutritional and energy problems. But man must be prudent and keep a critical and lucid eye on what may only appear to be progress.

M.S. How do you see the role of the doctor and "medical power" in the world of tomorrow?

H.L. That's a very big question. I believe of course that the doctor has power; but what worries me is not medical power as you see it. I'm thinking of an interview I recently granted to *Le Monde* in which I used the following comparison. A mechanic in his coveralls puts the wheel back on my car. If he has repaired it badly, I run a good risk of having an accident that could be fatal. Thus, he has power over me to the extent that my life depends upon the quality of his work. The surgeon who removes a stomach also holds the patient's life in his hands. He can very easily kill him, and thus he has very great power. . . . But we don't speak of the "power of the mechanic"; we talk about "medical power." Why? Although the mechanic's power is real, society does not recognize it. He is not an important personage who is greeted when he passes by. But it's a different matter with the surgeon, who is greeted with deference, who may run for election to the Senate or the City Council. In short, he enjoys a standing recognized by the whole of society. I say that we should question all false powers. I believe that our system rewards with its esteem individuals who work for its furtherance. I do not grant to a graduate of the Ecole Polytechnique, *a priori,* any social power. He is only a cog in the overall society, rewarded because he is more useful to it, and because the ultimate aim of society is productivity. If the ultimate aim were happiness, there would be no reason for rewarding him.

M.S.   The Supreme Court of the United States can revise or even reverse a decision made by the executive or by Congress to the extent that it seems harmful or contrary to the interests of society. Couldn't we have, based on that model, a supreme court competent in the field of health care? Health, the most precious of our riches, could thus be preserved despite certain baneful measures that might be adopted in haste.

H.L.   That court would be nothing other than a committee of sages. Such institutions abound in France. I can tell you that all you'd find on those committees would be the same "important personages" I just mentioned. Inevitably their mentalities, their prejudices are stamped on the decisions they make. You'll never see a teacher or a mechanic on one of those committees; and yet they have as much good sense, if not more, than all the mandarins of science. They merely have other sources of information. In any case, such committees will change nothing. It is time for man to transform himself and sweep away the value judgments on which civilizations have been built for 10,000 years. I'm radical in what I have to say. It's not a question of improving the system we have chosen in our so-called democratic manner. We have to invent a completely new way. I'm optimistic, because man *can* learn to transform himself.

M.S.   For that matter, we may *have* to, because I'm sure that the economic crisis is only going to get worse. The crisis is deepening, and we must learn to live more austerely if not in greater poverty. The medical power structure can help us to be "economic in health." Can one conceive of a preventive medicine that is not coercive? Could we ban alcohol and tobacco? In any case, life in society is based on the imposition of constraints. But how far can we go in the area of medical practice?

H.L. A preventive medicine for all would be possible if the end of each individual were integrated into that of the species; that is, if there were no subsystem that could be privileged at the expense of others. Once an entire population is vaccinated, you can very well risk not getting vaccinated. You have nothing to fear. That example, which is a bit too simplistic, can be extended to other realms. There will always be individuals who will find good reasons for not getting in line, for not being subjected to coercion. In that event, they must be convinced with arguments that are sound, not just emotional. The important thing is to offer convincing, substantiated proofs that prevention is indispensable. Prevention as it is conceived today is necessarily bureaucratic, constraining, and coercive. It is coercive in that people say: "We technocrats know what is good for you. Have faith in us." And people notice that the technocrats are wrong, that they make bad mistakes because their views grow out of self-interest, and then no one trusts them anymore. Preventive medicine should be social medicine—medicine that builds new human relations.

One of my latest books is called *Inhibition of Action*. In that book I show, with serious, substantiated arguments, that all pathology depends upon the inhibition of action, and that our modern societies increase the inhibiting factors every day. Inhibition of action in the search for pleasure leads to the destruction of biological equilibrium. A new structure of our inter-individual, inter-group, and international relations, a social approach to health, should be the basis for the preventive medicine of tomorrow. It would no longer be coercive, and would have no need to be, because it would not be *imposed* on people. It would allow for a broader approach to man in his environment.

M.S. In other words, health would be happiness and not the contrary, as the popular adage goes.

H.L. If you like, yes, that's the idea. But I would add that you won't see it happen any day soon.

JACQUES ATTALI
Medicine Under Prosecution

"*Ein Wunderkind,*" the Germans would say—a child prodigy. At less than forty years of age, Jacques Attali is an economist of international reputation, a professor, a political adviser much heeded by the Socialist Party, and a versatile writer—the author not only of theoretical works in his own discipline but of noted essays in such various fields as politics, music, and, recently, medicine. The book that he published in the fall of 1979, *The Cannibalistic Order or the Rise and Fall of Medicine,* refueled the debate in France not only about the therapeutic act but about all the existential problems, from birth to death, that underlie medical care in the West.

What makes Attali run?

For his friends, so much energy expended in so many directions at once is disconcerting. For his enemies—and he has many of them, because of his political opinions, despite his amiable personality—this very gifted man is suspect. Rooted in the soil of reason, of measure, of the *"juste milieu,"* the establishment has always been distrustful of intellectuals.

Jacques Attali, with his excesses, his outrageousness, his constant, feverish questioning, is no doubt disturbing. But in these times of crisis, don't we do well to be more "bothered" than reassured?

M.S. Why have you, an economist, taken such a passionate interest in medicine, in health?

J.A. In studying the general economic problems of Western society, I found out that health costs are among the main factors in the economic crisis. The production and maintenance of consumers costs a lot—even more than the production of the commodities they consume. People are produced by the services they render one another, especially in the field of health, where economic productivity is not growing very fast. The "productivity of the production of machines" is growing more rapidly than the relative productivity of the production of consumers. That contradiction will be eliminated as health and educational systems become more commercialized and industrialized. One look at our economic history and it's easy to see that our society is, more than ever, transforming craft activities into industrial activities, and that a growing number of services are becoming mechanized.

The confluence of these two questions leads one to ask: Can medical care, too, be produced by machines that will one day replace the doctor?

M.S. That question seems a bit academic, theoretical. . . .

J.A. Of course. But it goes a long way toward explaining the present crisis. If medical care could, like education, be mass-produced, the economic crisis would soon be resolved. This is somewhat the viewpoint of the astronomer who says: "If my reasoning is good, there is a star there." If this reasoning is accurate, and if our society is coherent, the logical conclusion is that, just as other functions have been devoured by the industrial apparatus in the earlier phases of the crisis, so medicine is becoming a mass-produced activity, which leads to the metaphor.

Doctors are being replaced by prosthetic devices whose role it is to repair bodily function, restore it, or take its place. If the prosthesis tries to do those things, it behaves as the organs of the body do, hence becoming a copy of one of the body's organs or functions. Such devices would thus be objects destined to be consumed. In economic language, the metaphor is clear: it's cannibalism. The body is consumed. Beginning with this metaphor (and I've always believed it was the source of knowledge) I asked myself two questions. First, is cannibalism a possible form of treatment? Second, does there exist a constant in the different social structures such that an accepted kind of cannibalism, dissociated from one's experience of it and reduced to the lowest common denominator, would be found again in therapeutic behavior?

First, cannibalism can be seen, on a wide scale, as a basic therapeutic strategy. Second, it seems that all strategies for healing a disease consist of a series of operations carried out by the body itself but also by cannibalism, and that one finds in all these strategies the following: selecting the signs that one is going to observe; monitoring them; denouncing what is going to break the order of those signs, what one calls Evil; negotiating with Evil, separating Evil from the rest. All healing systems employ these operations: selecting the signs, denouncing the Evil, watching, negotiating, separating. These different operations are equally applicable to political strategy: selecting the signs to be observed; watching them closely; denouncing the Evil, the scapegoat, the enemy; and driving him away. There are very profound connections between the strategy followed to combat an individual Evil and the strategy employed against a social Evil. This is what made me think, basically, that the distinction between social Evil and the individual Evil was not a very clear one. These various fundamental operations apply to different historical periods, to different conceptions of disease, of Evil, of power, of death, of life, and thus of what identifies the Evil—and effects the separation. In other words, the operations and the roles are the same, but the actors who play those roles are different. And the play does not always last the same length of time.

M.S.   From that to a theory founded on historical or mythical cannibalism. . . . Your essay upset and shocked not just doctors but the patients that we all are, potentially. In short, public opinion. . . .

J.A.   That essay tries to do three things. First, to recount an economic history of Evil—a history of its bearing on disease. Second, to show that there are, in a way, four dominant periods and hence three great crises between which the see-sawings of the system are structured, and that each see-saw motion affects not only the healer but the very conception of life, death, and disease. Third, to show that these see-saw motions concern the signs and not the strategy, which remains cannibalistic, and that in fact we begin with cannibalism only to return to it. In short, industrial history can be interpreted as a machine for translating basic cannibalism—the first relation to Evil, wherein people eat people—into industrial cannibalism, where people become commodities that eat commodities. Industrial society would appear to function like a dictionary going through three different stages of translation and thus resulting in intermediate languages—in a sense, four major languages. First there is the basic order, the cannibalistic order. It is here that we find the first gods appearing as cannibals; and in the myths that follow, historically, the cannibal gods eat one another. Then it becomes frightful for gods to be cannibals.

In all the myths I have studied, within different civilizations, religion serves to destroy cannibalism. For cannibalism, Evil is the souls of the dead. If I want to separate the souls of the dead from the dead, I must eat the bodies—because the best way of separating the dead from their souls is to eat their bodies. Thus, the idea of separation is fundamental to cannibalistic consumption. That's the point I wanted to make: Consumption is separation. Cannibalism is a formidable healing force for the power structure. Then why isn't cannibalism practiced anymore? (What I am about to say is evident in the myths. And in my essay I put forth an interpretation both of Girard's work on violence and of Freud's *Totem and Taboo*, in which he sees the totem and the totemic meal as basic, with the totemic meal disappearing into sexuality.) Well, from the moment I say that eating the dead permits me to live, I'll find some to eat. Thus, cannibalism is healing, but it also leads to violence. And it's in this way that I try to interpret the transition to sexual taboos, which are always the same as cannibalistic taboos. Because it's evident that if I kill my father or my mother or my children, I'm going to stop the reproduction of the group. And yet they are the easiest to kill, since they live next to me. Sexual taboos are secondary to food taboos.

Next comes ritualization, the religious dramatization of cannibalism. In a sense, one delegates, represents, sets the scene. Religious civilization is a dramatization of cannibalism. The signs one observes are those of the gods. Illness is possession by the gods. The only sicknesses one can observe and cure are those of possession. Healing, finally, is expelling the Evil—and the Evil, in this case, is the Devil; that is, the gods. And the principal healer is the priest. There are always two healers on duty. The denouncer of Evil and the separater—people we will later encounter under the guises of physician and surgeon. The denouncer of Evil is the priest, and the separater is the practitioner.

On the one hand, I tried to show that Christian ritual is basically cannibalistic. The texts of St. Luke on "the bread and wine," which are "the body and blood of Christ"—and which, if one eats them, give life—are cannibalistic texts and, of course, therapeutic. There is a medical, and at the same time cannibalistic, reading of those texts that is fascinating.

I next tried to recount the history of the Church's relation to healing, and to show little by little—around the twelfth and thirteenth centuries—the emergence of a new system of signs. Illnesses come not only from the gods but from the bodies of humans. Why? Because the economic organization is beginning. People are emerging from slavery. The dominant diseases are epidemics, which begin to circulate like men and commodities. The bodies of the poor transmit disease, and correlation between poverty (which didn't exist before, because almost everyone was either slave or seigneur) and disease is absolute. From the thirteenth to the nineteenth century, to be poor or sick meant the same thing. Hence, the political strategy with regard to the poor and the sick was the same. When one was poor, one got sick. And when one was sick, one got poor. Disease and poverty did not yet exist. What did exist was to be poor and sick. And once the poor or sick man was designated, good strategy consisted in separating him from others, containing him, not healing him but destroying him. In French texts, this was called "confining"—*enfermement* in Foucault's vocabulary. People were confined in various ways: the quarantine camp, the lazaretto, the hospital, and, in England, the workhouse. The Poor Law and charity were not means of helping people but means of designating them as such, and containing them. Charity was merely a form of denunciation.

M.S. The policeman took the place of the priest as therapist.

J.A. That's right. Religion withdrew and assumed power elsewhere because it could no longer claim the power of healing. Of course there were already doctors, but their role was limited to providing consolation; for proof of this, we have only to remember government authorities, very astutely, still did not recognize the doctors' diplomas. The political power structure considered its principal therapist the policeman, not the doctor. For that matter, in the Europe of the time, there was only one doctor for every 100,000 people.

But now I come to the third period, when it was no longer possible to confine the poor because they were too numerous. They had, on the contrary, to be supported and maintained because they had become workers. And so they stopped being bodies and became machines. The signs one observed were those of machines. Illness, Evil, took the form of a breakdown. Clinical language isolated and objectified the Evil to an even greater extent. Thus, Evil was designated, separated, and expelled.

During the entire nineteenth century, with public hygiene as a new means of control, the new binds of repairs, and the new distinction between doctor and surgeon, the policeman and the priest were replaced by the doctor.

M.S. And today it's the doctor's turn to fall into the trap. . . .

J.A. Today, the crisis is threefold. On the one hand, as in the preceding period, the system can no longer assure its own proper functioning. Today, for the most part, medicine is incapable of treating all diseases because it costs too much.

On the other hand, there has been a loss of faith in the doctor. People have much more faith in quantified data than in the doctor.

Finally, we witness the emergence of diseases and forms of behavior that no longer respond to the methods of classical medicine. These three characteristics lead to a kind of natural contiuum that moves from clinical medicine to prosthesis. And I have tried to set forth the three overlapping phases in that transformation.

In the first phase, the system tries to endure by monitoring its financial cost. But that leads to the necessity of monitoring behavior and hence of defining norms of health and activity to which the individual must adhere. Thus the notion of an economical profile of a healthy life.

From that, we go on to the second phase, which is that of self-diagnosis of illness (which corresponds to the designation of Evil) thanks to the tools of behavioral self-control. The individual can thus conform to the norm and become autonomous with respect to his illness.

The principal criterion of behavior was, in the first order, to give meaning to death; in the second order, to contain death; in the third order, to increase the hope of life; and in the fourth, that in which we live, it's the search for an economical profile of a healthy life.

The third phase is marked by the appearance of prosthetic devices that make it possible to designate the illness (Evil) in an industrial context. Thus, for example, electronic medication such as the pill coupled with a microcomputer makes it possible to release in the body, at regular intervals, regulating substances.

M.S. In short, health care, with the appearance of these electronic prosthetic devices, will be the new driving force of industrial expansion. . . .

J.A. Yes. In conclusion, all the traditional concepts disappear: production and consumption disappear, life and death disappear, because the prosthetic device makes death a fluid event. . . .

I believe that the important thing in life will no longer be to work but to be in a position to consume—to be a consumer among other machines of consumption. The dominant social science up to the present has been the science of machines. Marx is a clinician because he designates the illness—the capitalist class—and eliminates it. In a way, he says the same thing as Pasteur. The dominant social science of the future will be the science of codes—data processing plus genetics. My book is a book about codes, because I try to show that there are successive codes: the religious code, the police code, the thermodynamic code, and today the data-processing code and what is called sociobiology.

M.S. Does your thesis lead to a concrete approach to medicine, even in the long run? Does it constitute the beginnings of concrete ideas by an economist and politician on the organization of the medical profession and medical practice?

J.A. I don't know. For the moment, I don't want to ask myself that question. I believe that the first thing I wanted to show—the only thing—was that healing is a process in full evolution toward a model of organization that has nothing to do with the present one, and that we have a choice between three types of attitude: to preserve the practice of medicine as we have known it; to accept its evolution and see that it is the best possible, ensuring equal access to prosthetic devices; or to rethink our view of illness entirely, in order to arrive at an acceptance of death and an awareness that the urgent thing is not to forget or delay or await death but to want life to be as free as possible. I think that people will eventually have to choose from among these three solutions; and I want to show that, in my opinion, the last one is the truly human choice.

M.S. That's social utopianism. It's sometimes dangerous to be utopian. . . .

J.A. Utopianism can take two different directions, depending on whether we are talking about utopia as a dream of an absolute, in which case the dream is one of eternity, or as something that has never taken place in which case we then try to see which type of utopia is most likely to be achieved. I believe that if we want to understand the problem of health care, we must realize that there are realizable utopias. The future is necessarily a utopia; and it's very important to understand that it need not be fraught with danger, because to talk about utopia means to accept the idea that the future has nothing to do with merely continuing present tendencies.

I would even say that all futures are possible but one: the continuation of the present situation.

M.S. Is the future you postulate one in which a whole panoply of drugs will help man tolerate his condition?

J.A. I'm frightened by the fascination with drugs that reduce anxiety. People are trying to find ways to make anxiety bearable instead of trying to learn how to stop feeling anxious.

The medications of the future that are tied to behavior control could lead to political difficulties. It might be possible in fact to reconcile parliamentary democracy with totalitarianism. For totalitarianism to take hold, we would need only to maintain all the formal rules of parliamentary democracy but at the same time to generalize the use of those drugs.

M.S.   Does that seem possible—an Orwellian 1984 based on a pharmacology that would control behavior?

J.A.   I don't believe in the Orwellian model of technical totalitarianism with its visible and centralized Big Brother. I believe, instead, in an implicit totalitarianism with an invisible and decentralized Big Brother. Those machines that keep watch on our health, that we could use to our good, will enslave us for our good. In a way, we will undergo a gentle but permanent conditioning.

M.S.   How do you envision twenty-first century man?

J.A.   I believe that we must make a very clear distinction between two kinds of twenty-first-century man: the twenty-first-century man of the rich countries, and the twenty-first-century man of the poor countries. The former will certainly be a man much more anguished than he is today, but he will find the answer to the pain of living in passive flight, in antipain machines and antianxiety machines, in drugs; and he will try to live a commercialized form of the good life, no matter what the price.

But I am convinced that the great majority of people, who will know about machines and life-style of the rich but will not have access to them, will be very aggressive and violent. From that distortion will arise enormous chaos, which will be expressed either by racial wars or by the immigration into our countries of millions of people who want to share our way of life.

M.S.   Do you believe that genetic engineering is one of the keys to the future?

J.A.   I believe that in the next twenty years genetic engineering will be as banal, well known, and commonplace procedure as the internal combustion engine is today. The analogy is, in fact, particularly apt.

The internal combusion engine presented us with two options: either to favor public transportation and facilitate people's lives, or to produce automobiles—tools of aggressiveness, of consumption, of individualization, of solitude, of stockpiling, of desire, of rivalry. . . . The second option was chosen. I believe that genetic engineering occasions the same kind of choice, and that unfortunately the second option will again be chosen. In other words, genetic engineering could pretty much create conditions under which humanity could either take responsibility for itself freely but collectively, or else devise a new commodity, genetic this time, made up of copies of people sold to people, of chimeras or hybrids used as slaves, robots . . .

M.S. Is it possible and desirable to live 120 years?

J.A. Medically, I know nothing about it. I've always been told that it is possible. Is it desirable? First, I believe that the industrial system in which we find ourselves no longer sees an increase in life expectancy as a desirable objective. Why? Because increasing life expectancy only makes sense if the human machine's threshold of profitability is similarly increased. But as soon as a person gets to be older than sixty or sixty-five, and his productivity and profitability begin to slip, he costs society dearly.

Hence, I believe that the very logic of the industrial society will require that the objective no longer be to prolong life expectancy but to see to it that man live in the best way possible—but with health care expenses as reduced as possible for the sake of the collective. Then we witness the emergence of a new criterion for life expectancy: the value attributed to extending life expectancy will not be as great as that placed on maximizing the number of years a person lives without illness, and particularly without hospitalization. Actually, from the viewpoint of the cost to society, it is much preferable that the human machine abruptly stop functioning than that it deteriorate very gradually.

This is perfectly clear if we remember that two-thirds of all health-care expenses are incurred during the last months of life. Likewise, all cynicism aside, health-care expenses would not reach a third of the present level (175 billion francs, or about $35 billion, in 1979) if people all died in automobile accidents. We have to recognize that logic no longer resides merely in increasing life expectancy but rather in increasing life expectancy *without illness*. I think, however, that expanding life expectancy remains a fantasy that serves two purposes, the first of which is mainly a question of self-preservation of the power elite. The more totalitarian or centralized societies tend to be run by "old" men, and are in fact "gerontocracies." Secondly, capitalist society hopes to make old age economically profitable by making old men solvent. Right now the elderly are a "market," but not a solvent one.

This all fits in very neatly with the view that man today is no longer important as a worker but as a consumer (because he is replaced by machines in the workplace). We see very well how the big pharmaceutical companies operate today in relatively egalitarian countries where retirement is adequately financed. They take aim at their target and favor geriatrics, at the expense of other fields of pharmacological research, such as tropical diseases. Thus, the technology of retirement determines the acceptability of increasing life expectancy.

For my part, as a socialist, I am against the increasing life expectancy, because it's a decoy—a false problem. I believe that posing this type of problem enables us to avoid more essential questions such as how we go about freeing our time in the present. What is the use of living 100 years if all we gain is twenty years of dictatorship?

M.S. The world to come, "liberal" or "socialist," will need a revamped, "biological" morality—an ethical code to cover cloning or euthanasia, for example.

J.A. Euthanasia will be one of the essential instruments of future societies. Socialist logic is based on freedom, and the exercise of the most basic freedom is suicide. The right to commit suicide, directly or indirectly is an absolute value in this type of society. In a capitalist society, machines for killing, prosthetic devices that make it possible to eliminate life when it has become too unbearable or too expensive to sustain, will be used routinely. Euthanasia, whether an expression of freedom or a commodity, will be one of the givens of the future.

M.S. Will the citizens of tomorrow be conditioned by psychotropic drugs and subjected to manipulations of the psyche? How can we guard against this?

J.A. The best way to protect ourselves is to educate ourselves and increase our scientific store of knowledge. We will have to ban a great number of drugs. But perhaps the point of no return has already been passed. . . .

Isn't television, for that matter, an abused drug?

Hasn't alcohol always been an abused drug?

The worst drug is the absence of culture. People want drugs because they have no culture. Why do they seek alienation by means of drugs? Because they have become aware of their impotence, their inability to live, and that impotence is expressed concretely in a total *refusal* of life.

An optimistic bet on man would be to say that if man had culture, in the sense of tools for thinking, he would be able to escape solutions that only aggravate and deepen that impotence. To grasp this Evil by the root is to give people a formidable instrument of subversion and creativity.

I don't believe that the banning of drugs will suffice. If we don't attack the problem at its root, we shall inevitably become enmeshed in the machinery of the police state, and that's worse.

M.S. How are we going to handle mental illness in the future?

J.A. The evolution of medical practice as regards mental illness will occur in two phases. In the first phase we will still rely on drugs, psychotropic ones, which have meant real progress in the treatment of mental illness during the past thirty years. In the second phase, and for economic reasons, we will begin to rely more on electronic means of treatment—either to control pain (biofeedback), or provoke psychoanalytic dialogues (a data-processing system). This will then lead to what I call "the explicitation of the normal." That is, the electronic apparatus will make it possible to define the normal with precision, and to quantify social behavior, which will then become economically consumable because the means and criteria for conformity to norms will exist. In the long run, once a given mental illness is conquered, the temptation to conform to a "biological norm" will condition the functioning of social organization.

Medicine reveals to us the evolution of a society that will orient itself toward a decentralized totalitarianism. The desire, conscious or unconscious, to conform as much as possible to social norms is nothing new.

M.S. Will forced normalization govern all the realms of life, including sexuality, since science now makes possible the almost total dissociation of sexuality from conception?

J.A. I think that we will go very far in that direction. The production of people is not yet a market like any other. But following the logic of my general reasoning, I can't see why procreation should not become one. The family, or the women, thus becomes a means of production of a particular object, the child. One can, in a way, imagine "wombs for rent"—something already possible, technically. This notion corresponds perfectly to an economic development in which the woman or the couple will take part in the division of labor and in general production, making it possible for people to buy children as they buy peanuts or a television set.

If, on the economic plane, the child is a commodity like any other, society will in turn consider it such, but for social reasons. The survival of the collective depends upon a sufficient pool of people. If, for economic reasons, a family does not want to have more than two children, the interests of the collective will be at risk. Thus, we get an absolute contradiction between the interests of the family and those of society. The only way to solve the contradiction is to allow society to *buy* children from the family. I'm not referring to family subsidies, which are feeble incentives. I mean that a family would agree to have lots of children if the state would guarantee it both the payment of progressive substantial allowances and specific reimbursement of all material expenditures for each child. Under such a plan, the child would become a kind of medium of exchange between the individual and the collective.

What I have just said is not something I take lightly or view complaisantly. It's a warning. I believe that the world we are building will be so frightful that it will mean the death of humankind. So we have to be prepared to resist it; and it seems to me today that the best way to do so is to understand and engage in the battle in order to avoid the worst. That's why I take my reasoning as far as I can.

M.S. Resist what, since you foresee a world of prosthetic devices?

J.A. The prosthetic devices I forsee are not mechanical but will be used to combat chronic afflictions linked to tissue degeneration. Cellular engineering, genetic engineering, and cloning are preparing the way for the development of prosthetic devices that will in effect replace defective organs.

M.S. The increasing role of data processing in society calls for a revaluation of ethics. Do you see this increased reliance on data processing as a threat to man's freedom?

J.A. It is clear that all the talk about preventive medicine, the economics of health care, and good medical practice will make it necessary that each person have his or her medical record on tape. For epidemiological reasons, all such dossiers will be centralized in a computer to which doctors will have access. The question arises: Will the police have access to those records too? Sweden has this kind of sophisticated system but does not have a dictatorship while certain other countries do not have this record-keeping system but *do* have dictatorships. In the face of new threats, we must know how to create new procedures. Democracy has a duty to adapt to technical evolution. The combination of old constitutions and new technologies may lead to totalitarian systems.

M.S. One of the commonest predictions for the future is that man will be able to exercise biological control over his own body thanks to microprocessors, among other things.

J.A. That control, which already exists, in the form of pacemakers for the heart, and likewise for the pancreas, should ideally be extended to apply to the elimination or reduction of pain. Researchers predict the perfection of little implants capable of releasing, in the target organs, hormones and active substances. If one's aim is to prolong life, such progress is inevitable.

M.S.   It seems that we are leaving an era of physics to enter an era of biology—something close to a panbiology. Do you agree?

J.A.   I believe that we are leaving a world controlled by energy to enter a world of information. If matter is energy, life is information. That's why the major product of tomorrow's society will be living matter. Thanks in particular to genetic engineering, new therapeutic, nutritional, and energy tools will be developed.

M.S.   What is the future of medicine and medical power?

J.A.   In a rather brutal way, I would say that just as washerwomen have been displaced by advertising images of washing machines, so doctors integrated into the industrial system will become the developers of biological prostheses. The doctor as we know him will disappear, yielding his place to a new social category living off the prosthetics industry: inventors, salesmen, installers, and repairmen—much as exist now to keep those washing machines in running order. What I have to say may seem surprising. But I wonder how many people are aware that the main enterprises interested in prosthetic devices are the big automobile manufacturers like Renault, General Motors, and Ford.

M.S.   In other words, we'll no longer have any need for internists because "normalization" will be effected by a kind of preventive medicine, self-managed or no, but in any case "controlled." But how can we accomplish this without resorting to force?

J.A.   The appearance on the market of implements for medical self-monitoring will create a preventive-medicine mentality. People will adapt to conformity to the norm. Preventive measures will not have to be forcibly imposed; people will embrace them to achieve social acceptance. But we must not lose sight of the fact that the most important thing is not technological progress but the highest form of commerce among men, represented by culture. The shape society takes in the future will be a function of our capacity to master technological progress. Will we dominate it, or be dominated by it? That's the question.

ELIE SHNEOUR
The Deeds of Mother Nature

ELIE SHNEOUR was on the faculty of several universities, is an advisor to the United States government, and is president of a society of researchers. A biochemist with a special interest in neurology, his reputation is worldwide, particularly for his work on the effects of infantile malnutrition on the development of mental capacity.

M.S. What are your views on the future of medicine: What is "do-able" and what utopian?

E.S. A reasonable projection of the gamut of tomorrow's medical methods is that they will be dramatically changed but in ways we can already foresee. There will probably be a shift from the concern about molecular diseases—diseases that have a chemical or biochemical origin—to those based on life-style. The medical profession will probably be unrecognizable by the year 2000. What is utopian is that one might be able to do away with disease altogether, and that genetic selection will be such that only those labeled perfectly healthy persons, if that can be defined, will survive and carry on the race. Of course, we allow people to survive and to live who one hundred years ago would never have survived, and these people are increasing what some people claim is a genetic load on the population: It is becoming increasingly unable to adapt to changing situations. What is utopian is to find a way to do this effectively and maintain a population that will be able to survive and grow and develop under those conditions. I don't believe that this kind of genetic selection will happen. I think that the whole concept of what disease is and what health is, will become a significant factor. So what is utopian is that we will do away with disease altogether, as it is defined today.

M.S. Isn't utopia particularly dangerous in the realm of health?

E.S. I don't think that utopia is dangerous at all: it's so far from realization that it won't have any significant influence.

M.S. It can give false hopes, illusory expectations.

E. S. I don't think so. I think that technology has already given the world many false hopes. We have expected that if man could go to the moon we could cure cancer, and we have discovered that it is not the case. Of course the viewpoint of the average human being, the middle-class individual, is that if you can solve the problem of going to the moon then you can solve every other social and technical problem. Technology has had until now an almost undisputed position in the world, especially in the Western world. Slowly we are beginning to recognize that there are limitations, that technology cannot solve all problems. It's easier to solve technological difficulties than sociological problems. I don't think that the hope for a utopia of health is believed any more because technology is losing some of its credibility.

M.S. Do you foresee any major medical breakthrough in the fifty years to come, or before the year 2000?

E.S. I think the major breakthrough that will come in the next fifty years, by far the most important one, will be the understanding of how information is processed by a cell. That is the fundamental thing that underlies practically all health and disease problems. All biochemists, all medical doctors look at a living cell of any sort—whether it be bacterial, animal, plant or human cell—as an energy-transforming entity that can convert one form of energy into another: it metabolizes food and produces wastes that are removed. This is a mistake. I think the major thing we will recognize is that this is not the purpose but the consequence of the fact that the cell is first of all not an energy-transforming unit but an *information-processing unit.* Everything it does is based on its ability to utilize information that starts from its immunological memory, the immunological consciousness. The physiological consciousness in the human being, or memory, the ability to recognize a chemical friend or foe is a consequence of the cell's ability to process information. The consequences are what the cell does. When you look at an electronic microprocessor, an element of a computer, it's very small; the battery for the electricity it needs is about five or ten times as large. So if you never knew what a microprocessor was and you had the combination of a source of energy, the battery, and the microprocessor in the corner, you would look at the battery and look at its energy contact and completely miss the microprocessor. But the microprocessor is the reason for having the battery, not the other way around. We look at the cell, we look at health and disease, we look at the battery because it's very visible. It's big and it's fairly evident and obvious, but we don't look at the reason for having the battery in the first place, which is the information transfer from the element that is a microprocessor. All we know about the cell is ATP, the conversion by adenosine triphosphate into energy, the formation of complex proteins, and so on. The disease process is based on the ability of the cell to "recognize" normality, to recognize the steady state, the standard health process, and the pathological process. Before you can do anything about it, you must be able to recognize it; then the response may exist. This opens up many possibilities, the replacement of a limb for example. A person may lose an arm or a leg or an organ. This organ may be replaced one day by making use of the information processes contained in DNA. The most important breakthrough will be the ability to define as well as to control the information-processing devices in a cell or in a combination of cells, to do what you want it to do, to be replaced in case they are damaged or sick. All the methods that we use today, like protheses, are very crude methods. We are really, in effect, doing things to the battery, not to the microprocessor. What I am looking at is the thing that is responsible for making an effective working unit, and that is the microprocessor part. Of course I don't want to imply by this that the cell is like a computer. It isn't. But I am using this as a very poor analogy because there is no comparison between a computer and the ability of a cell to do things. If we could, by using all the technology in existence today, match the ability of a cell to process information, the equivalent for one single cell would occupy probably the volume of a building. And, in the same analogy, if we could make a machine matching the human brain, which is very small, this machine would be of the order of the size of the highest skyscrapers in New York. The idea I want to point out, is that information is the key to the future.

The ability to understand and to control what the cell or a combination of cells can do, and to control the mechanism by which information is processed and transformed, would be the most extraordinary advance in the next fifty years. All the rest is minor by comparison.

M.S. Do you see twenty-first century man as less aggressive, or will the demographic explosion, the greater density of urban living, the greater scarcity of natural resources, make him still more aggressive?

E.S. I think that the obvious statement to make is that there will be more people, problems will be more complex, the density of human population will increase and there will be a decrease in natural resources. If that is true, and probably for several years it will be true, men will have to become more aggressive. There may be mechanisms and medical advances that will make him less aggressive, but this would be a palliative, not therapy, to allow him to survive and to adapt himself to a situation that is basically not acceptable. But I would say that that will probably not be the case in the long term. I think that in the next fifty years the most important thing that will happen, from the point of view of protection against aggression and all other factors that are the result of the problems into which the question enters, will be that there will be somehow a drastic decrease in population, in one form or another, either through disease or through wars or through simply reaching the asymptote of what a human population can stand. I think that the human being simply was not designed for and is not able to operate effectively in the existing Western industrial environment. He was never meant to be that way, and to the extent that he is able to survive, it is because he has surrounded himself with palliative methods to reduce the tension, to make himself less aware of it. The best example is the heavy use of tranquilizers, which don't cause man to revolt or try to do something about the environment but, on the contrary, to fit into it and to fit it by making him less aware, by affecting his consciousness. I think this is a bad way to learn. I think that fifty years from now we will look back at today and say, "How could they have survived?" and two things will happen. We will have to reduce the population but by reducing population we will probably increase lifespan very considerably, maybe to one hundred and fifty or two hundred years. If the Bible is any reflection of the past, there is no reason to doubt that it was possible at one time to live to one hundred, two hundred, three hundred years of age.

M.S. Most scientists agree, more or less, that even with "good genes," the maximum span of human life would be 120 years.

E.S. I'm not sure that you can say that. We know that single cells have survived a long period of time. In Egyptian tombs bacteria have been found that were several thousand years old and if you put them in a proper medium they will multiply. It is true that there is a mechanism in each cell, that we are probably programmed genetically; there is an aging gene, genetic aging instructions: after a certain age is reached, the mechanism begins to break down. There is evidence of young children who aged rapidly. For example, by the time they were less than a year and a half old, they showed evidence of being geriatric. There is growing evidence that there is a mechanism in which—if you remember my answer to the previous question—the information processing may be modified and even controlled. So a person theoretically should be able to live an infinite period of time by replacing organs. There may be a mechanism by which human beings will be selected: those who have the right combination of genes will be allowed to survive while the others will not. I visualize a smaller population eventually—maybe in a few hundred years—who could live forever, theoretically. They would live except for possible destruction by natural catastrophe, being hit by a meteorite or something, but theoretically a substantial part of the population could live an infinite life. I'll go even further. I'd say that the fundamental drive for modern technology, the ultimate motive for its development, is for man to control his destiny. But much more than that, the conquest of death is probably the ultimate reason for all of this: trying to maintain health, to have children to perpetuate oneself, to write books, to leave something beyond natural life. The idea is to try to avoid death altogether. Therefore the ultimate aim may be to conquer death, and I would say further, to conquer time.

M.S. Does genetic engineering promise a Golden Age or an apocalypse?

E.S. It promises both. It is impossible to see any technological or scientific advance that will allow men control without looking at both sides of the coin. Take a hammer, a plain hammer such as was used by Michelangelo to make the famous sculpture La Pièta. The same type of hammer was used by a crazy man to try to destroy La Pièta. You know the story that Szent-Györgyi told me, about vitamin C? One day I said to him, "Surely there is no danger in vitamin C. Ascorbic acid cannot possibly do people any harm—your discovery does not have a reverse side." He said, "Oh, no, I am a murderer. I killed millions of people." After World War II he visited Germany and saw a warehouse full of vitamin C, ascorbic acid, and he asked his host what it was for.

The Deeds of Mother Nature

They said they were able to send soldiers to the Russian front for years without citrus fruit and were also able to send them on ships and submarines; otherwise the men would have died of scurvy, and the war would have probably ended one year earlier. Szent-Györgyi tells me that during that one year more than a million people died and that he feels responsible for those deaths. So any technological advance has two sides; the question as it is posed is really not meaningful. Genetic engineering has a marvelous positive side: the use of information processing of the cell. The control of technology is as important as the proper use of technology.

M.S. The fear of genetic engineering seems to be replacing the fear of an atomic holocaust—an unmanageable, uncontrollable technology.

E.S. I think you're quite right. Sooner or later these terrors will be expressed in some form: there might be an atomic holocaust or there might be genetic control so that some members of the population will be able to survive at the expense of others. All this is quite possible, but there is something else going on, the contrary mechanism, a very powerful force that is growing, and that is communication between people. It's very difficult to realize today that about a century and a half ago, in the United States, there were people who were crossing the continent in covered wagons with horses. Imagine even a little meeting like we are having now, a discussion in the middle of Omaha at night, around a camp fire, one person saying, "You know, one hundred years from now there will be flying objects carrying people in a few hours from one coast to the other when it takes us months." He would have been laughed out of the place. Or the concept of the telephone, or television. Information is now transmitted extremely rapidly, and I think that's why there has been no use of atomic warfare since World War II. If there had not been communication of the speed that exists today, there would have been nuclear weapons used in Vietnam and elsewhere, no doubt. So my feeling is that information, technology has created a monster or potential monsters—I'd like to think that it's not really a monster—but it has also created mechanisms to protect mankind, and the most powerful mechanism is information.

M.S. Do you mean to say that today, with recombinant DNA, we are in the same position as mankind before the release of the bomb?

E.S. The difference between the two is in the time scale. The consequence of an atomic bomb took place in a very short time span. The total effort—from the moment the first reactor went critical under the stadium in the University of Chicago under Enrico Fermi's direction in 1942, until the bomb was dropped in 1945—is a very short span of time. And don't forget that most of the basic technology was already known. In the case of genetics, let's assume that tomorrow you wanted to start to do genetic engineering. This cannot be done on a massive scale by putting some stuff in the water you drink to cause genetic change. The remarkable thing about biological systems is that in spite of their apparent vulnerability they are so tremendously resilient, they are stronger than the strongest metal, the hardest rock, the highest mountain. They have the ability to remove mechanisms that might destroy them. Any kind of genetic engineering has to be done on a one-to-one basis; consequently you can produce one human being or two human beings or a hundred, then there will be a generation gap before they are reproduced. Thus the damage you could do is limited. It's questionable whether bacteria can be created that could be more devastating. In fact the whole argument about DNA recombinants today is that nature itself has created "monsters" over the years. People have said that nature is not hostile. Nature can be very hostile. It has created the most dangerous bacteria—anthrax, plague—and it is difficult to imagine that it could be possible to create an organism more virulent or more deadly than those which nature has produced. Nature has had much longer experience and is able to do much more severe genetic damage than a human being could ever do. The time it would take to develop such a thing would, in my opinion, prevent this kind of holocaust, the apocalypse concept, by analogy with atomic power.

M.S. I believe Oscar Wilde said that nature imitates man, but in this case man can hardly try to imitate nature.

E.S. I don't think that man is capable of being as cruel, as devastating as nature, and that is the one saving grace. In spite of his arrogance, his amazing arrogance—even in the Bible, in Genesis man is said to have dominion over all things that are alive and dead on the earth—his arrogance is much greater than his ability to perform. He cannot make landslides or thunderstorms as well as nature can.

M.S. Is living to 120 years possible? Is it desirable?

E.S. Yes, definitely. I just turned fifty, and I found it very traumatic. If I knew when I was a twenty year old what I know today, how much better I could have done things. And I think of when I will be seventy, and if I knew at fifty what I'll know at seventy, how much better and so on. Death obliterates all the experience accumulated in a human being's life and the only way we can convey it is by writing something down: passing information. That is why books are written, paintings painted, movies made, things done. It's to produce a record of one's experience. But the funny thing is that it doesn't do any good. People knew how terrible wars were, but that didn't stop wars; how terrible atomic energy is, but that won't stop it. You tell your kids. I tell my children not to do something when they are twenty. Then they'll do it, and they shouldn't have done it. But anything I could do or say wouldn't make any difference. They have to experience it themselves. There is a level of experience that cannot be taught—it must be experienced. And then when we have got it, it is already too late, you are going to die.

With advanced technology it would be possible to live 120 or even 150 years and to live effectively, to make use of the information. This will have a great influence on society, with extraordinary consequences. If I could live to be 120, I would be much more concerned about politics and the future of mankind. For example, there is not going to be any petroleum in 20, 30 years from now. I might say that I don't give a damn, because I won't be alive by then. But if I was going to live 120 years or 150 years, I would give a damn. The consequence of an effective prolongation of human life span may be extraordinarily positive and effective, because it would make people concerned. And, as a result of that, ameliorate the situation politically and otherwise. People would take much more effort and initiative to see to it that by the time they got on to 120 years old, things would be better. I think it was Churchill who said that the human life span of memory lasts five years. Things that happen within five years are very vivid, but things that are beyond five years are part of history. It has been shown that when you go back to Beowulf and the old writings, like the *Chanson de Roland*, you find, if you read them carefully, that the span of thought of the past is one year, maybe two years. So I suspect that as life span grows, memory span of immediacy becomes increasingly large. And as life becomes increasingly longer and you give a damn much longer, then you begin to do things that may, as a group, as a society, improve the chance of survival of the collective memory ten, or fifteen years beyond.

I can give you an example of something far more immediate: look at the preamble of the major social documents of the Western world, starting with the Congress of Vienna. The 1815 text of the agreement of the Congress of Vienna begins, "We the Count of Talleyrand for France, etc., do hereby agree that" and you see a very narrow span of single individuals, not only time span but also the tiny size of the social unit. You read the 1919 Treaty of Versailles, and the group is larger. It's no longer the Count of Talleyrand for France, it's "we the high contracting parties." But the United Nations charter begins, "We, the People."

There is progress; there is some recognition. First it was a very narrow egoistical view, and that seems to be broadening. Whether it is true or not, that's not the point. It moves in the direction of the greater social unit and the greater time span, both of which cannot but be positive from the standpoint of society.

M.S. Will euthanasia pose problems requiring the forging of a new morality?

E.S. Yes. I think that one should have the right to control one's life span, and I think that the reason suicide has such a negative impact goes back to the ancient tribal concept. All modern religions started at the tribal level. And at that level, the individual person plays a very important role for survival. And if a person decides to take his own life, it may affect the rest of the tribe in its struggle for survival. All our religious concepts, all our morality, biblical, Judeo-Christian morality, is based on the "farming" morality, rather than the hunter morality. The settlement of man, civilization is associated with farming, not with hunting. Man was first the hunter, then he became a farmer, and to make the transition acceptable we had to change the rules. "Thou shalt not kill" is meaningless from the hunter's point of view, or even "thou shalt not steal," because there is nothing to steal in his way of life. So morality is really a function of the society in which man lives. We have to re-examine our morality for the twentieth or the twenty first century angle.

M.S. That will be a biological morality.

E.S. Exactly, that's including euthanasia. In the future, people will not use such crude things as taking poison, or shooting themselves in the head. I think that there should be ways and means to help you. Instead of being in a hospital for instance, you will give a big party at home before you die. At the end of the party, you take a potion that will make you feel very good until you die.

M.S. The Socratic way?

E.S. Well, Socrates was in rather severe pain. Hemlock is a pretty unpleasant death. But there are the drugs that the American Indians had for euthanasia, there is some evidence for that. We have drugs that can make you euphoric: you listen to good music and have a chat with your friends until you are finished. Our modern society has made such a tragedy of death because we don't know how to handle it. The old society knew how to handle death, Eastern civilizations too. We don't know how to handle it. We surround it; we close the door on people who are dying. We remove them from being part of society, which is one of the cruelest things we can do. We put them in the hospital with all kinds of nasty equipment on them. We give them some cold technology and we forget them.

M.S. As we do with old people.

E.S. That's so. Old people have great wisdom. And that's why, in isolating them, we commit a social error of enormous implication from the point of view of medicine. We have become nuclear families. In all major cities, in all urban environments the old people live in one place, the families with children live in another place, the singles in a third place. In the past, the old cities, medieval cities, cities until the nineteenth century and early twentieth century were so vital because people lived together. It's why today we have so many divorces, so many problems. There is no stability at the family level because they are separated instead of having the wisdom and the experience of several generations in the same house: the grandparents influencing the parents, providing the balance for the children so they see continuity. What we lack in our society is continuity. A sense of continuity, which comes back to the information transfer, information processing, not only on the biological level but also at the social level.

M.S. Will man's free will and liberty be endangered by the new psychotropic drugs? How can a free society arm itself against manipulations of the psyche?

E.S. I'm not so sure that man has freedom of choice. I don't think it's the drugs that have alienated him. These psychotropic drugs simply confirm the fact that he is alienated. The question should be put the other way around.

I don't think that man would use psychotropic drugs unless he found a need for them; before they existed he used alcohol, a very potent psychotropic, or tobacco or other drugs. The essential purpose of these drugs is to change man's perception of himself and of his environment. The reason that he does this is necessity. By the time he is tense, he reaches for a Valium pill or a drink: I don't think that his liberty or freedom of choice has been alienated. The alienation already exists; he takes the drug so that he can become less conscious of it.

M.S. In the framework of our society, the abuse of psychotropic drugs may become a way to manipulate people. The Nazi regime was a very totalitarian state; it was only slightly tempted to use drugs as a social means of control, maybe because the psychotropes were not, at that time, so sophisticated. Later, in the Soviet Union, psychotropes became a way of governing: the use and the abuse of drugs on tens of thousands of people. Can we conceive that a government—by putting some drug in the drinking water, I don't know how—can manipulate the psychoses of citizens with pychotropic drugs? The range of psychotropes is broader and broader, their effect is more and more complex and subtle: not only sleeping pills or drugs to give you a state of euphoria but nearly all aspects of life may then be controlled. Sex, for instance, is going to be enhanced or reduced drastically, and so on. So I foresee a big brother society where all people can be manipulated.

E.S. I think it's possible to do some of this in a fairly limited way and within well-circumscribed geographic zones. I don't think it can be done on a national or international scale, for the same reason that poison gas could not be used effectively: the difficulty of controlling the extent of its activity. I think that if you try to put something in water, then you must know the exact limitation of where the water goes, what it is used for, how long it will be used and what will be the effect over a period of time. The same goes if you put the drug in food or something else. I think that the trouble with the "manipulators" is that they provide no effective control. You can get them started, but you can't necessarily stop them or you cannot start them exactly when you want to. That, in my opinion, is impossible to handle. In order to make the drugs useful as a means of control, the thing that you want to control people with must itself be controlled. I don't believe that there are mechanisms, either that I can think of now or I can imagine in the future, that would make possible this kind of effective control. There are other ways to do it in a totalitarian state, so I don't think it's a problem.

M.S. You think that what the Russians are doing now is the maximum that one can do? To take ten, twenty, or fifty thousand dissenters, people who are considered potential enemies, put them in asylums, and try to "neutralize" them with "punitive" psychotropic drugs.

E.S. Absolutely. But I don't think that you can do this on a broad scale, because then it could boomerang. As I said at the beginning, technology has a very strong limiting function that people don't always realize: whatever you do will boomerang. You must always look for the boomerang before you estimate the control established by the technology.

M.S. Can mental illness be helped by psychic manipulation, psychotropic drugs, or electronic means?

E.S. I think that psychic manipulation on an individual basis has a large range of possibilities that will become increasingly effective and significant in years to come and will pose very important ethical problems. The question of what is mental illness is also an extremely important question. Very few people deny the fact that there is mental illness. But if one person in a room goes into a panic because he thinks an earthquake will happen and he has all the evidence for it and everybody else thinks that he is crazy, he might be put away, even if he is the most alert and knows what is going on. So how do you determine a person's mental illness? The determination of who is mentally ill is one of the most pernicious determinations that one can make. The mechanism of deciding is one of the most pressing social problems we face. Who in fact is mentally ill?

M.S. One can roughly say a person of the present society who is not adapted to it and escapes it, taking refuge in an inner world.

E.S. I'm not even sure that this is correct. I know people who are highly maladapted by every possible standard, but what they say makes extremely good sense—if you assume that certain things are going to happen or are about to happen. I think that certain criteria can be used. How do you determine mental pathology? Number one—the probable versus the possible. If a person is worried about possible things he is sick, but if he is worried about probable things he is okay. The distinction between what is probable and what is possible is one distinction one can make. The other one is consistency. If a person has a point of view that is consistent, he may not necessarily be ill, but his premises may be wrong. The premises may have been brought about by a variety of experiences in the past that may be totally at variance with what is going on, but that does not necessarily make him mentally ill. His premises may be different from yours or mine. Consistency and probability versus possibility are in my opinion the bases for any kind of judgment. But one has to be very careful about how one defines these words because of the enormous potential damage one can do our social structure, if we believe that man is free or likely to be free.

M.S. From the very narrow limits of what we really know about mental illness today, the deep pathology, what progress do you envision?

E.S. These will be cured. These will be solved. Breakthroughs are not a dream of the future; they are happening right now: by means of drugs, by electronic means, by surgery. The severest psychotropic diseases recognized—pathological diseases such as schizophrenia, which appear to have a molecular basis—are on the path to being cured. Of course, there is not just one type of schizophrenia, there are probably dozens. They may have basically the same symptoms, but they have a variety of related but different causes. But still, I think that most of them are likely to be cured or are curable within the next decade or two. Many laboratories are looking at some of these questions in a completely new way, which suggests that some clues already have been found that are likely to result in significant breakthroughs eventually. I foresee effective treatments by various means, probably chemical, biochemical, or immunogenetical methods. Not electronic means or surgery necessarily, because they are too crude a method.

M.S. What will sexuality be like in the future? In a world that has contraceptive vaccines, will sexuality be completely dissociated from conception?

E.S. I think that conception and sexuality have nearly always been dissociated. In primitive societies there was no idea that the sex act had anything to do with conception; the time span of nine months is too long to relate one to the other. It's only in modern times that we have made the jump. The churches, the Roman Catholic Church in particular, and before them, of course, the Jews, have been very effective in making that link, and it's caused serious moral problems. But I think that the two have been totally disassociated and that the period of history when they were associated is very small compared to the span of human prehistory and history.

M.S. That's part of the new morality, the new biology we are talking about versus, in this case, Judeo-Christian puritanism.

E.S. Right, that's part of the new biological morality. Contraception is a tough problem because the creation of life, conception, first appears to be the easiest thing in the world to interfere with. And then it's discovered that it's one of the most difficult things to interfere with. Nature has some extremely ingenious ways to avoid being interfered with, and it's not only ingenious, it's cruel. For example, if you interfere with it, the pleasure level goes down. Many contraceptive methods involve interference with the pleasure of the sex act; to dissociate the two would be the greatest boon of all, but even the pill has side effects that interfere, so this is a joke that nature plays on humanity. Something that looks so simple to interfere with—to get pure pleasure at the expense of conception—but actually the two are very difficult to separate. I will say that this is a problem with enormously greater implications than we realize. The world of tomorrow may not only dissociate conception and sexuality but may actually put conception under state control. I think the days are coming when you can have a child only by permission of the state.

M.S. As in China, today.

E.S. The case of Singapore has even greater significance, because it is not as coercive, in a way. In Singapore, if you have one or two children you get a tax advantage. After that you have to pay for the full support of society for having children. And the economic factor is already playing a role. But I think much more important than that, biological methods will be used in which you will be permitted to have children, and if you have children without permission there will be very severe social and economic consequences—either a very heavy fine or forced abortion or a variety of other things. The bearing of children will be increasingly controlled by society. I'm convinced of that. So that you will have to completely separate sexuality from conception, not only biologically but politically.

M.S. Can you imagine a world in which, thanks to prostheses and organ banks, parts of the body could be replaced like automobile parts now? What ethical considerations would this raise?

E.S. Definitely. I think you will be able to replace every organ but not through organ banks, you know, taking an eye, taking a cornea or a heart or a kidney. These are temporary methods that are very crude. I can see that by using genetic engineering it will be possible within the next twenty or thirty years to begin to create organs—either growing them internally, or generating them outside the body and grafting them—but they will be created from your own cells, using your own genetic material. A mucous membrane cell will be taken; the remarkable thing about all living cells of a human being is that they are totipotent. Every cell has the complete biological DNA information of every other cell. It's all simply being selective: some switches have been turned off, and other switches have been turned on. By turning on all the switches in the right sequence, you should be able to convert the cell into any organ you want to, if you find the proper medium around it for this particular organ to grow. And assuming you have kidney failure, you will be put on a machine for six months or a year. During that time one of your cells will grow into a kidney. A few months later, when the kidney is grown, it will be installed in you. The only organ that you won't be able to replace is the human brain, because that is your own personality. In fact, I can see the day where you can buy a brand-new body: this is a way a person might live as long as several hundred years. When you finally reach a point when there are too many things wrong with the skin—scars, whatever—anyway you want to be taller or you want to go in a different direction, you want a body that is more appropriate to the things you want to do, you can look in a catalog, select. There will be advisors, professionals, for a certain fee, they will determine what kind of body would be most appropriate for what you want to do. They will provide the right specifications, use one of your cells and I can see tragedies when they find that the brain cells are different from the rest of the cells, that many of them won't be able to find the right cells, and they won't be able to replace the body once in a while. But you find a cell that is totipotent, grow it, and eventually a body made to your specifications will be grown by using genetic engineering: adding a gene, or removing one gene from that cell to provide you with a completely new body. But you won't be able to replace the brain. That is the unique thing about the human being, that won't be able to be changed, without changing the person.

Ethical considerations exist only when we have to take an organ from another human being. There's a problem, unfortunately, that will be with us for probably several decades: the definition of death. When can you actually take an organ? Under what conditions? That is a severe ethical problem. There are very few kidney machines, who decides who lives and who dies? Will it be a poor woman, who is fifty or sixty years old, who doesn't have a high school education, will it be her or the Nobel Laureate who is dying of uremia because of kidney damage? Will they select him instead of her? Then everything we say about equality before the law or before death will no longer play a role. They will let the poor woman die and select the Nobel Laureate for survival. Some of these questions will be resolved by the fact that there won't be any need for kidney machines. I visualize within the next five to ten years that you will be able to carry a little instrument—like you now can carry a pacemaker—that will clean your blood outside, an outside electronic kidney through which the blood will circulate. I think this is well within the range of possibilities and will probably be one of the first things that will happen. A machine has already been developed that is one tenth the size of the usual kidney machine and could be run by an individual. So that ethical problem will be resolved. There will be others of the same kind resolved the same way. But until then we will be faced with the problem of who decides.

M.S. We have already, I think it's in America, in Seattle, a kind of "selection" committee.

E.S. Oh, there are committees all over the country, not only in Seattle. A committee that makes the decisions exists in every major city. It's done right here in San Diego.

M.S. And who makes the decisions? What kind of body is this?

E.S. Its made up of different groups, some are physicians, others clergymen, or public figures. Problems are looked at, decisions are made. The decisions must be made; somebody must make them. If there are only six kidney machines and seven people, one person simply will have to do without. The question is who is that to be. It's a decision that you don't want to make but that you have to make. As on the battlefield. There are six people here and three will definitely die. You've got to let them die and you have to try to save the other three. The concept of triage is inevitable when the resources are limited. Somebody has to make the decision, and it's never fair. What you want to try to do is to make the least unfair decision possible.

M.S. Either because of a population explosion or scarcity of natural resources, will sizeable human communities exist in space or under the sea?

E.S. I think so, but people are a bit too optimistic on that for the following reasons. It's possible to live in space or under the seas or in the seas in good-size communities, but I suspect that the psychological problems of living in a totally foreign environment are likely to be enormous. I can visualize putting colonies on the moon to mine things and for energy purposes or to install a telescope to see the stars, or a launching pad for the universe beyond that. That I can see, but I cannot visualize human communities outside the earth. Maybe my children and others will be able to, but I can't. Don't forget I was associated with the space program for years, and I'm not only aware of it, I contributed to that. But the only reason why this may happen is the exhaustion of natural resources that would only occur because of large population pressure. If you want to ask me the worst single human problem that humanity faces today, I will say rapidly rising population. But the most serious technological threat is nuclear warfare. All the problems we face—war, misuse of technology, exhaustion of natural resources—result from that very large population. I suspect that we won't be affected by any plague or any other major diseases, which had a very positive effect, mind you. After the black plagues from 1346 to 1353, Europe's population was wiped out by 50 percent. But that made the Renaissance possible. There were fewer people; the pressures were less. You didn't have to wait ten years for your boss to die so that you could get his job; the farmer could take over the farm earlier; it was a very positive thing. I'm not even sure whether if it hadn't been for 1918 that Europe could have recovered as quickly as it did following World War I. I think many of the problems we face today and as a result of World War II—which to a great extent were brought by technology—are the result of huge population pressure. All the major social problems we face today, every one, including disease, comes right back down to the population increase, which is enormous. I think the first thing that will happen in the future—by mechanisms that I cannot predict—will be a reduction of population. I'm convinced of that, and I hope it will not be done by man. That is to say, by war, indiscriminately. This is a very important factor, because war selects against the best people. In France in World War I, one and a half million of the best young Frenchmen were killed, and probably the vigor of the French nation was sapped to this day because some of its best genetic material was wiped out. On the other hand disease, even cruel disease like the plague and the pandemic picks the people that are the least resistant, and the selection is more positive.

It's a cynical view, but, you'll remember that humanity deals in statistics. We as individuals are very concerned about our own lives but survival of societies depends on statistical selection of the best or the most effective members. I'm an elitist in that sense, but a biological elitist. Of course by saying that I don't mean that a strong man who wins the Olympics is necessarily the best choice. It may be a man who can hardly walk but has a fantastic brain and good eyesight. I don't know what the right selection is, but nature knows better than we do, I think.

Nature doesn't give a damn really about the individual. It cares about the statistical assembly, the probability in broad terms. Nature is a mathematical game. Something that of course Einstein refused to accept. He said that nature doesn't play dice with destiny. But I think that was one of the few times Einstein was wrong.

M.S. Will immunotherapy someday palliate the failings of chemotherapy and surgery?

E.S. Yes, I think, it will. I look at these as way stations on the way to a major city. These are temporary mechanisms to solve local problems. I think that therapy will be one of the least important functions of immunology. Back to the first thing we discussed, the fact that the biological system is primarily an information processing system. Immunology will provide means to make that information more effective. I have no doubt that sooner or later immunology will give us the possibility to get rid of the most aggressive therapeutics, in the very narrow field of illness. There is no question that that will be the case. But I don't like the word palliate. Palliative means it will cover up but not really act as a therapeutic. It will do more than that.

If you said not paliate but replace chemotherapy, no question. I think most surgery and most chemotherapy are very crude, barbaric methods of solving problems. Certainly in cancer, this is true. It's the most barbaric approach. When we talk about progress in cancer and say we have better surgery, this borders on the criminal, but we can't do better. We have to live with it for another few years. I'm not even sure of that. There is some strong question, for example today about the radical mastectomy for breast cancer. Will it in fact increase a woman's life span after the operation? There is no evidence that this is the case. In some cases it may very well be. In many cases a more conservative operation will give the woman much less disfigurement and the chance to live just as long. So I definitely agree that immunology and of course genetic engineering will do a tremendous amount of good.

In fact I would say that surgery is the ultimate proof of the failure of therapy, except in cases of traumatic surgery, repairing organs in the case of accidents. Chemotherapy is not quite as bad, because it is possible to have selective chemotherapy that is much more specific, that goes to the target organ. But the classical chemotherapy, a massive impregnation, means inundating the whole body without control; not only the target organ but everything else has to suffer along with it. We will soon have improvements in chemotherapy, where you can go directly to the target organ. It will not only be because of the delivery system but also the form of the chemistry. You use a chemical element that goes everywhere, and one of the few molecules out of the billions circulating in the blood—let's say one tenth of them—will go where they are supposed to do the job. The others are totally wasted or are doing harm. I can visualize the individual molecule being packaged with a chemical coat so that it will go everywhere but it will do nothing until it recognizes the target organ and then it will be released. That is already coming into being, in fact our own group is working on some aspects of that.

M.S. And then it's related to immunology.

E.S. Not necessarily. The recognition may not be an immunological one. For example a propos the generation of insulin by the Islets of Langerhans in the pancreas, the identification of insulin *in situ* will cause a chemical change in the molecule. The molecule is packaged in such a way that the presence of a hormone or, for example, uric acid, or reatine phosphate in the blood stream, any of these compounds can trigger a release of the active principle to the organ but not to any others. Nature does that all the time. Prolactin is converted to lactin in the stomach, you have enzymes, proteolytic enzymes, which are not active until they are triggered, they are made into an active form. The same thing is true of the very complex mechanism of blood clotting. The platelets break under certain physical conditions, causing the release of the appropriate substances. I think that we can mimic many of these things. Nature does it all the time. There are plenty of molecules circulating in the bloodstream and the therapeutic form or the effective form is released only on the site; there is already enough evidence to do many of these things. This way you can reduce the side effects of such drugs; I think that is a very important future aspect of modern chemotherapy.

M.S. Isn't computer-controlled public health a prelude to an even more police-like conception of the society of tomorrow, one that would not have any possible escape?

E.S. I don't know what you mean by computer-controlled public health. If you mean biological information in the file of the citizen as a means of identification as well as prevention, (HLA and so on) yes, indeed. I think this is one of the most frightening aspects of information processing.

M.S. At the same time it is indispensable, or very useful.

E.S. I'm not sure of that. I visualize a system in which each person will have individual computer data, and when you go to the doctor or the specialist at the hospital the data is placed in the computer when needed. That record may eventually be carried in your own body, maybe even under the skin. Small storage, all the basic information. In fact, it doesn't even need to be taken off the skin: it can be inserted in the skin. Whenever you go to the physician, you put your hand against a machine, which will read your data, so that it's not available anywhere else.

M.S. So what's the difference? If the police check you, they check your hand instead of looking in their computerized files.

E.S. I know, but it's quite different because first they can look in the file without your being aware of it, and secondly, these machines are likely to be sufficiently complex and limited so that they cannot be available to the police or government but only to the health authorities. The health authorities would not have the records of your health itself. The only records are kept in your own person. Perhaps a copy would be available in your house in case you get hit by a car or something like that, but I can visualize a system in which all the advantages of modern technology can be made available without having a central authority getting all the records. I think you can carry the record with you and make it available only when needed, and those machines will be available not to the police but to the hospital, controlled by hospitals, and the medical health care authorities.

It's a problem, but this is the way to solve it eventually, and that is true of many kinds of things, schooling, everything. Instead of an identity card you must show to people, you carry this information on yourself. The police will only have certain kinds of information available from this. Their machine, for example, would tell them whether you were arrested before, things like that, whether you were drunk or whether you are reliable. So that each will have a machine that can only unlock the information to which they are entitled and no other. I think you can use modern technology in a decentralized way. In the Soviet Union you have a central file. I visualize decentralization in the democratic countries. Decentralization is going to become one of the most important single factors in proper use of technology. Not the "small is beautiful," idea, but that the social unit will become smaller. A smaller unit will have control. A small group will have information about you but not the larger one. That information would have to be made available by you, and under your control; I think that's possible and desirable. There will be a new Bill of Rights. We talk about the Bill of Rights of today, which was written 200 years ago. But there will be another Bill of Rights and another *Déclaration des Droits de l'Homme* for the electronic and the computer age, which will include the limits to which people can have information. It's part of the new morality, including biological morality.

M.S. Can man exercise biological control over his own body by the use of a miniature apparatus, made possible by minicomputers?

E.S. The answer is yes. I can foresee the time when it will be possible to control your blood pressure and adrenalin level. If you want to run, do some exercise, or you're about ready to drown, by means of a mechanism you can create additional adrenalin to save yourself. In other words the human body can be adapted and trained to do a remarkable number of things, including running a four-minute mile, which is a remarkable achievement, a physiological achievement controlled by the brain. For many years, everybody tried to run one mile in four minutes. It was not successfully accomplished until that young physician in England, a medical student, did it. It was the control of mind over matter in the sense that he was able to train himself to do something absolutely remarkable without the use of outside agents. It is not desirable to multiply these gadgets because they bypass the mechanism of evolution and the mechanism by which a living organism can do remarkable things but also restrain and refrain. I believe that living human beings can do remarkable things, and we have not even begun to tackle them. The human being himself has not made full use of his own body, his brain, his physiological strength. We live in an age that I think is part of the coming biological morality. Until now, we have had very little faith in the body's ability to survive. The whole health care system is based on that. When you have a cold you go to a doctor, whenever you have a headache you take an aspirin. We always have to do things from the outside without enough faith in the ability of the human body to do something. I don't want to go as far as Coué or the Christian Scientists. I don't think that by wishing it, you can stop appendicitis. The unhappiest person is a Christian Scientist with appendicitis; there are limits beyond which the body needs outside help. But I think that the vast amount of potential is never used because we have no faith, and I think part of the coming biological morality is the return to faith in the human body and a recognition of our potential, an examination of what it can do. And all of the other mechanisms? I would like them to build airplanes, to travel long distances quickly, that kind of thing. But when it comes down to the control of the individual, I'm not sure I'm in favor of it because eventually we will be so mechanized and so controlled that we will devote most of our time to worrying about how we are, a kind of medical introspection. We live in an introspective age where everything is me, my, myself, what's happening to me and so on, and disregard of the fact that man is overall a social animal. So I look at it with a great deal of concern. I would like to see man return to being confident, to the idea that the human body properly trained and developed, early in life especially, can do remarkable things.

 M.S. Those gadgets are already a growing industry.

 E.S. I think that eventually these gadgets will damage if not destroy independence, individual consciousness, and free will. What is the use of living? Eventually the point will be reached when we will be all the same: whenever we feel ill, we'll push a button. I'm not so sure that Chopin would have written the great music he did if he didn't have tuberculosis. In fact, there is good evidence that he thought he was going to die, so he had to write music; he had terrible coughing and to forget it he used to write music. Beethoven wrote the ninth symphony when he was deaf, perhaps to overcome the limitations of his infirmity. Roosevelt became president not because of but in spite of the fact that he had infantile paralysis. You go back through the history of men who have accomplished something for themselves and for humanity, they have done it not because but in spite of the odds. The challenge of these odds is an essential part of the creative process and gives energy and purpose to life. Maybe I am very old fashioned and people will laugh at me a hundred years from now. Still, I don't think so. I think that challenge gives purpose to life, and the main problem of our society today is the lack of finality. Men have their cars and their nice houses and their swimming pools and they go on vacation for two weeks and they work. But they have lost purpose. Why do they do all these things? What is the meaning of all this? The overcoming of odds is an essential part of survival, and I mean long-term physiological as well as psychological survival. I worry a great deal about this problem of autodiagnostic devices, which become automedication by proxy.

M.S. How do you envisage the role of the doctor or medical power in the future?

E.S. The doctor will lose a considerable amount of his power for two reasons. The first is information. Information will be widespread. There will be not one kind of doctor but maybe fifty different kinds of doctor, in all fields, so that the power of the doctor will be diluted. And if we go along the lines of what I discussed a moment ago, namely, that the potential of the living organism is much greater than we give it credit for, the influence of the miracle-dealing physician will become less important. The physician will become a part of the health care industry. He will provide therapeutics—treatment within some narrow, circumscribed area—and achieve limited improvement, and therefore will cease to incarnate "medical power." More important than medical power will be health power, the definition of health, who can have certain jobs, who cannot, and so on. How long a person is going to be able to keep his job, things like that. I see the role of the physician as becoming more important but less powerful.

M.S. Can you conceive of preventive medicine that is not coercive?

E.S. The only preventive medicine that is not coercive is generated by the human being himself. The only kind of preventive medicine we can talk about is preventing lung cancer, for example, by refraining from smoking and things like that. I'm not so sure that you can predict what is going to happen with other kinds of things. Right now, how many cases do you know of presidents of the United States, such as Lyndon Johnson, who had a marvellous checkup and then a heart attack shortly thereafter? Or Eisenhower, who had the best medical care that money could buy, developing a serious intestinal disease, illeitis, soon after. What kind of preventive medicine could have stopped that? I think that short-term preventive medicine is meaningless. Long-term preventive medicine—developing certain good habits and certain dietary conditions—may play a role. I think the most important role, if you want to live a long time, is to select your parents very carefully. The genetic basis plays a major role, and environment has to be adapted to the genetic potential of an individual human being or society. It's very difficult to select your parents; that's the only problem about it.

JONAS SALK
A Stoic of Our Time

IT IS EVIDENT for those who are paying attention today to the leading research carried out in biology that medicine of the eighties will be to medicine of the past what the pocket calculator is to the huge computer of yesteryear. Isn't medicine that calls forth a defense response—as opposed to today's medicine, which is mainly intervention therapy, often aggressive—the panacea, the future, the recourse to the natural equilibrium that good health should be?

Perhaps the most renowned of all living immunologists is Jonas Salk, whose works on poliomyelitis vaccines made him famous all over the world, but whose philosophical works—deeply rooted in Eastern wisdom—are often overlooked. From the East, Jonas Salk has brought back not only abstract concepts but also a way, a style of living—both rich and austere—which is flourishing both at the institute bearing his name (a masterpiece of California architecture: shadow and lighting effects, a labyrinth of cubes, as strict as a Japanese "Zen" garden) and at his home. What a refuge, this house in La Jolla—hanging over an abyss, an unspoiled inlet of the Pacific Ocean—with books and bright paintings on all the walls. It is in this house that Jonas Salk welcomes a few visitors with the stand-offish serenity of a guru.

M.S. I'd like you to give me a few ideas about what you consider to be utopia in medicine.

J.S. What do you mean, utopia? Are you talking from the point of view of the physician, the point of view of the patient, the point of view of the scientist? How would *you* answer the question? Answer it first, and I might then understand what you have in mind.

M.S. Maybe I need a utopia just to help me face the future: in order to hope rather than be too despairing. I may need utopia in place of God, maybe.

J.S. Well, that's fine. I can understand that. When you ask me a question like that, I only think in realistic terms, more down to earth.

M.S. What you did was rather utopian, but of course, you did not know it was utopian. Let me put it in another way. When you were working on your vaccine, the future or the ideal for poliomyelitis was a more sophisticated system of prosthesis. Then, thanks to you and your work, what was a very complicated electronic problem (ways and means to get artificial arms and legs to work) just disappeared, vanished. By a simple biological means, it was over. So, the prospect changed completely from a kind of wild science fiction, into something apparently very simple.

J.S. So, what you are asking is to use the polio analogy, or the likelihood that there will be other diseases that might succumb in the same way. Diseases that might eventually be eradicated just as polio; eradicated from all parts of the world, with polio being eradicated from the world in the same way that smallpox has been. That will eventually make it, may make it unnecessary to immunize against polio, just as it is not necessary to immunize against smallpox. But the disease agent exists only in the human reservoir, and some means must be found to interrupt transmission from person to person. Now, there are some diseases that cannot be eradicated: tetanus, for instance. It will always be necessary to vaccinate against tetanus, because tetanus spores are ubiquitous. That's an example of a disease in which the dream would be the immunization of all children early in life, and if more children were immunized, tetanus would not be prevalent. So the important point for the public to understand, for a layman to understand, is what is possible and what is not possible. I believe that it is necessary now for the public to be educated as to what is and is not possible, so that

A Stoic of Our Time

they do not have dreams that have no basis in reality. You might turn around and ask me a question about cancer. And I will point out that essentially we know the cause of lung cancer, in terms of its relation to cigarette smoking. Now, what must be done to eliminate cancer caused by cigarette smoking? If I put that question to you as a layman, which you are not (but let's assume you are a layman), then you should be able to give as good an answer to that question as I do, because it is within your sphere. Now you might turn around and say, then why doesn't the scientist find some way to eliminate cancer by means of a pill, or by means of an injection, so that it could be prevented. Well, that's putting a great deal of stress on the system. I would then say, yes, but by what means can we prevent cancer due to chemical agents? And at the moment, we have none. That raises the general question of the control of cancer caused by chemical means or of the management of cancer, once it is identified. That moves us into another round of comprehension, so that if I am asked to discuss the dream of a world free of cancer, I cannot visualize a world free of cancer, because of the nature of the phenomenon of cancer itself.

M.S. Which we do not know anything about.

J.S. Well, on the contrary.

M.S. Well, we know bits and pieces, but not the basic induction mechanism.

J.S. We know a great deal about how cancers arise. Even the fact that some of them are linked to cigarette smoking, some are related to radiation. So we cannot say we know nothing about it. We may not be able to control these conditions, and we may not have a single specific etiological agent, as in the case of the microbe; nevertheless, we understand enough about neoplastic transformation—of cancer's transformation of normal cells—to recognize and realize that there are a wide variety of preconditions that result eventually in a cancerous transformation. Now, cancer cells arising from normal cells are so much like normal cell tissue that the immune system doesn't necessarily recognize the cancer cell and reject it as the immunological system should do. It is also likely that the cancer cell has a defense mechanism against the immune system, against its recognition, and in that way favors its own survival. Now, this much is known. It suggests that in order to deal with the cancer problem, all you have to do is eliminate the causative factor. In the opinion of the experts in the field, 80 to 85 percent of cancers are caused by chemical agents in the environment. Apart from lung cancer, some cancers may be caused by viruses, for instance leukemia, and in some cases, radiation may be the cause. Others are of genetic nature. But you see, with such diversity and predisposing causes as these, cancers are going to continue to arise in the population. And cancer is a different kind of disease than tuberculosis or smallpox or polio or measles. So, the dream of a world free of cancer is about as unrealistic as a world free of people.

M.S. Or free of death.

J.S. Yes, but death is particularly related to life itself.

M.S. Do you mean that there is reality, what is possible, and then the area between the actual and the possible, which is the realm of utopia or at least of the future?

J.S. To me, the meaning of the question is: "What do I visualize as the future of controlling cancerous diseases?" Well, there are going to be some surprises in the future related to the possible, where ways and means will be found to select or attack cancer cells when they appear. Therefore, it is conceivable that anticancer substances could be put in the water supply. These substances would selectively attack cancer cells and not normal cells. Those are kinds of science-fiction fantasies, which require so many steps between here and there; I prefer to think of the steps we can take to obtain necessary knowledge. And I don't encourage that kind of speculation, because it is like taking a drink or taking a fix of heroin each day, obliterating reality when there is so much that can be done here and now. Why engage in this kind of farcical fantasy when nothing is being done about the question of cigarette smoking and cancer now? That is the reaction I have to questions like this. Because I am aware of how much can be done to deal with problems of disease for which we have the knowledge, and about which nothing is being done, and I always turn questions like this around. I bring them into the present, pointing out how much disease there is still in the world, about which nothing is being done, and the various things that people expose themselves to, and the way they live, and what they do and do not do that interferes with their health.

M.S. Would you take a quick look at the predictions made some time ago by the Stanford Institute? It is funny to see, in two years' time, what seems crazy and what is almost certain to happen.

J.S. I wouldn't comment on them. What would be the sense of it? These are guesses. Anybody can do that, and I never participate in inquiries of that kind; that is why you may find me a not very cooperative subject in this kind of thing.

M.S. I see. Well, you know, in itself, it is quite amusing.

J.S. Most uninteresting to me.

M.S. It is not without any interest. Some scientists are reacting very positively, while some, like you, are very practical and say that's science fiction.

J.S. It is useless and pointless.

M.S. It gives hope; it's food for hope.

J.S. But I have so much hope in the "step-by-step" method that I don't need that. That's like a drug, as I said. It's not food; I prefer food for thought. I would rather explain the nature of some of these problems rather than simply make guesses like that. Then the reader could understand, by himself, the unreasonableness of this kind of prediction and would not even be interested in it. Then he would have some appreciation of the nature of the process and the phenomenon involved. I don't want to encourage this kind of pseudo-scientific journalism. I am critical of this cheap journalism; it's a pollution that contaminates and interferes with people learning. I would much rather see people fed with food for thought, as I said a moment ago, and be given a basis upon which they can understand what it is all about. Rather than sound like a magician and partake in this sort of thing, I do not even want to be associated with it.

M.S. What about man of the twenty-first century: will he live longer, better? Do you see him less aggressive, more convivial, or will the population explosion, the greater density of urban life, the greater scarcity of natural resources make him still more aggressive?

J.S. I see questions like that somewhat differently; I too ask what the future is, what does the future have in store in the sense of life expectancy? Do you think of life expectancy in terms of number of years? I like to think of life expectancy in terms of what can we expect life to be like in ten years' time, fifteen years from now, or beyond. And here, I take the view that there will be many evolutionary changes taking place, all of them governed by the needs that arise, whether the need has to do with the number of people in a given area and the resources necessary for their existence, or the amount of fuel they will need for their automobiles, whether that fuel or some alternative energy source is available. Each and every one of these factors will become the equivalent of evolutionary pressures, pressures to bring about evolutionary change. I believe that there will be a tendency on the part of young people, imaginative ones, to be challenged by the limitations and difficulties that exist in their lives and the lives of others. . . . Whether man is going to be more or less aggressive is a rather simplistic question because it's altogether possible that necessity will require less aggressive people. It depends on so many factors changing in small evolutionary steps and ever more rapidly that again, as in the other kinds of predictions, I think it could be projected and extrapolated from the present or the past, which does not mean that the future, the short-term future, will be the same as the past. There will be changes; one innovation will bring about another and so on, and I would hope these changes would be appropriate to meet the needs that arise. It's equally possible that people will have to learn how to get along much better and will, in a sense, mature in that respect, out of necessity.

M.S. That's an optimistic guess about human behavior.

J.S. Yes, if you like.

M.S. At least an optimistic view of man's behavior.

J.S. I don't share the views of the doomsdayers who make straight-line projections from Stone Age man to now to the future. I believe that people are trying to solve the problems that they have, and the tendency would be toward problem solving rather than problem creating. Of course, we still see barbarity in the world, bloodshed taking place, but I think it is different from what it was in the past, and it's going to be still different in the future.

M.S. Do you hold out the hope that people will share their wealth?

J.S. Out of necessity, yes. They won't offer to do it, but they will have to, out of necessity. Now, we have the United Nations, the United States of Europe if you like, whatever it's called—the Common Market countries—that was unthinkable in the past. And there is ample evidence in the Israeli-Egyptian accord of what will be necessary in the future. At the moment, the Arabs are acting up about oil, which is understandable; the countries that need oil will invent ways and means for not depending on that oil supply. It's obvious, but that's taking an evolutionary view. So, taking an evolutionary view and looking at the cancer problem and the other questions you posed, you end up with a reasonable basis for hope. And you don't terrify yourself with unreasonable fear or buoy yourself with unreasonable hope. I tried to explain both in terms of the mechanisms that advances offer: adaptive mechanisms in both cases up to the limit of adaptability. That does not mean that we won't have war, we won't have terrorism, we won't have violence, but we have to understand their causes and how we can best reduce them. In other words, reduce the amount of pathology in the world, whether physical or biological pathology, or psychosociopathological manifestations.

M.S. I know that you like Asian wisdom, you have been to India, you got the prestigious Nehru Award. In a way, I admire Indian wisdom and civilization. In another way, they are faulty: there is something repulsive in the fact that Indians are dying of hunger yet have one of the greatest concentrations of cattle in the world. So I am always a little bit suspicious of humanistic speeches not sustained by socioeconomic reality, as in India. How can you explain this Indian discrepancy between so much wisdom and so much poverty? Is this the price we have to pay for wisdom?

J.S. Not at all. We just have to put wisdom and practicality together. And not think of wisdom in the abstract, just as we must not think of these utopian questions in the abstract. All this must be brought down to earth, putting Indian wisdom together with Western practicality. Then you've got something. After all, West and East are going to meet; it doesn't make any difference in which direction. In time, some people will not survive (just as some animal species have disappeared) and will become extinct. Some philosophies, some systems of belief will not be appropriate under the circumstances that will prevail in the future. I see this as being an evolutionary process as well, being subjected to the test of evolutionary pressures. Now, my answer was in terms of what would be a more useful or advantageous formula: if man exists presently in the West or in the East, my response would be that I would hope there would be a meeting or a fusion or a hybridization of wisdom, on the one hand, and intelligence or intellect or practicality on the other. We see that in the United States, for example, where there is a great deal of interest in meditation and interest in Eastern philosophy in general, that interest on the part of the public has been publicized. When I was in India, I found that their points of view were extremely congenial and they found my way of thinking very congenial. I think that there are people in India who are close to the West in that sense. Those people lead very rich lives by virtue of having themselves done the hybridizing, which has not, however, gotten to the masses, nor has it transformed the country as a whole. You might say that it's just a matter of time before discoveries are made of modernizing or taking advantage of what we know about making life in the physical sense, in a purely biological sense, more agreeable. Now, the question is who is better off: those who live in the tribal villages in Africa, or those who live in downtown New York, or Los Angeles or some other places. So, there are tradeoffs we have to think about, and there is a tendency to criticize and find fault rather than trying to get the other person's side, seeing it from their point of view.

M.S. I would like to ask you about this genetic engineering business. There is a lot of controversy. Some see genetic engineering as the dawn of a Golden Age, while others see it as the beginning of apocalypse.

J.S. That is why I find the most useful way to deal with these problems is to explain rather than simply take sides, and then people understand what is meant by genetic engineering. Then they wouldn't have irrational nightmares or fantasies. We can already see that the terror that struck the public imagination when this first became public knowledge has all quieted down. Researchers are going about their business, and you will soon see the benefits of this technology. Personally, I do not see the harm that other people see. I think that, in a way, it is going to be equivalent in some respects, to the Industrial Revolution. It is going to be possible to produce valuable substances very economically. So, I hale it as the beginning of a great opportunity. Now, I see the opportunities and I see the advantages, and I do not see the danger.

M.S. Are you doing anything in this field? Do you work on recombinant DNA?

J.S. No, it's not my field. But you can be sure that you'll hear some interesting, positive things about it. People are afraid that we're going to create human monsters, but that's not even relevant to what the technology is all about.

M.S. There are fears about some frightful germs escaping biological control.

J.S. Number one, the technology is controllable; number two, that's highly unlikely. Let's assume that organisms will be created that can bypass existing antibiotics. But that can occur in nature, too. What was most feared about genetic engineering is exactly what happens in nature. We cannot play around with nature. We have to remain within the limits of nature's puzzle.

M.S. And we cannot invent new pieces?

J.S. Well, we can. In the sense that isotopes have been made, and we might be able to construct new genes. The elements already exist in nature: certainly the combination exists, too. And that is how evolution takes place. That is why we are here, in all of our complexity.

M.S. Is living 120 years possible? Is it desirable? Gerontology is a multi-billion-dollar business everywhere in the world. Do you think the incredible amount of money we are investing in this new science is worthwhile?

J.S. After all, people are going to live longer and they might just as well take advantage of that. Take advantage of the increased wisdom through experience that all people have. And do everything possible to keep themselves as healthy as possible, with the optimum functioning of their faculties. And I see what happens later in life as a continuum from the beginning of life. So the way to practice gerontology or practice pediatrics must consist of increasing our attention to all phases of human development from the beginning of life on. My approach to the problems of gerontology would be to see what should be done in the early years. I have heard people say jokingly that if they had known they were going to live as long as they have, they would have taken better care of themselves. That's just a way of saying that if we know people are going to live longer, we'd be able to prepare them for a longer life. And if people are going to live longer, then they have to recycle themselves and have more than one career.

M.S. That means that society has to provide the elderly with something to do. I've lived several years in the Far East and I think they have a better approach to old people than we have. Old people are part of the family there and are not rejected.

J.S. Yes, that would be part of my personal view of gerontology, to examine customs in other parts of the world, broadly, in terms of way of life, behavior, and attitudes. All these matters must be dealt with culturally as well as nutritionally and experimentally; what kind of supplements will be needed to enhance life in the same way that insulin enhances a diabetic's failing system. Whatever may be necessary to ameliorate defects or deficiencies that might arise. But this is just one minor aspect of the problem, although a necessary one. This problem belongs to the broad field of the human sciences and not only to biological sciences.

That is why I am opposed to early retirement, except in very special cases. I think that since people are going to live longer, and are living longer, we ought to develop some respect for the value that they can have and not retire them or force them to retire but take advantage of what they have to contribute. Something altogether different from what they did earlier in life. Therefore we have to figure out how we are going to take care of this valuable natural resource. Instead of looking upon older people as a burden only, look upon them in terms of the value they represent.

M.S. Well, retirement means death for many of them.

J.S. Indeed, And so we have a problem. The moment you create a class like that, you have created a problem for yourself. I have no doubt that this will have to change. As the number of elderly increases, the burden on the young to pay for the sustenance of the old will become unbearable. To answer these questions, I look at the mechanism or the dynamics of the processes involved. That is where we will get the clues and see the answers. I would rather see what route to take, what direction should our thinking take to integrate the future into the present or the present into the future. Old age, disease, death, must be integrated into the stream of life and be accepted with serenity.

M.S. Modern stoicism could, in a way, form the basis for a new morality, partially grounded on scientific facts. Could euthanasia, under social and political restraint, be part of a new morality?

J.S. I think we are going to have to come to terms with death, make it part of life and not look upon it as something either unexpected or undesirable.

M.S. Death is, after all, the most terrible injustice we have to suffer.

J.S. In a sense, it could be the most beautiful release, and relief. We may not want it to come if we want to extend our experience of life; there does, however, come a point in time when we may have had enough. And it may be that people will look upon having the option of death as a rather civilized thing to do. One might well ask the question: why do we have so many people on the face of the earth? And when we discourage births, we encourage the prolongation of life; we put all this out of balance. And life of what quality? When people are suffering and life serves no useful purpose for them or for anyone else, what would be or should be or might be a reasonable way of dealing with it? I am quite sure that, out of necessity, we will arrive at a much more reasonable attitude about this. Our present views will have to go through whatever the necessary or appropriate evolutionary changes are. Our attitude toward death and the way in which we deal with the problem of death become much more acceptable, much more reasonable and appropriate.

M.S. Speaking about a new morality shaped by the biologists always gets on my nerves. We are still living under the shadow of the Nazis' manipulation of human life; they too had a scientific and biological approach. So when biologists start talking about new ethics, I'm suspicious. Maybe in America, you don't have this feeling, but we have it in Europe.

J.S. My view is that biologists should be involved in ideologies, ethics, and metaphysics. As far as I'm concerned, it is a social question, a question that has to do with all of society. I don't know why biologists should be particularly excluded, since their science is a by-product of this society.

M.S. Because he views life and death even more than the priest or the funeral director.

J.S. Well, obviously biologists and physicians deal with life and they are the people who are specially concerned with optimal health and interested in it throughout its span. I am interested in the reduction or amelioration of diseases so as to allow health to take place. Now, at some point in time, the organism breaks down in body or mind. What do you do under those circumstances? Let us then narrow the focus to that part of life related to the fatal diseases, or the occurrence of accidents or injuries.

M.S. Now, you mean of course, management of terminal diseases, or the management of those people who have become "vegetables." Then it's a problem of euthanasia.

J.S. Well, it is either a question of euthanasia, a question of the leaving of life, or the continuation of life. Who should be involved in the decision in that regard?

M.S. It seems that in the state of California, you have some guidelines, provided by scientists, including yourself.

J.S. I don't think I was involved in that, though I think they are reasonable guidelines. I think the persons who are ill can request not to have any extreme measures applied to maintain their lives under certain conditions. In other words, they elect not to have their lives prolonged; they are allowed to die in a dignified, natural fashion rather than overextending their lives by artificial means. One has to consider what the most human attitude to adopt is, whether it is more humane to prolong life by artificial and costly means or to allow someone to die. It has to be dealt with on an individual basis, and I don't think one can generalize about this. I believe that, rather than legislate on matters of this kind, it should become a matter of custom. I think every physician will tell you that he, in one way or another, has participated in helping individuals at the end of their life. The whole thing has been overplayed and part of the overplay is frequently the effect of journalism.

M.S. Are you referring to Karen Quinlan's story?

J.S. Yes, that is exactly what I mean. I believe that the media (at least most of them) do the public a disservice by dealing with these problems in a superficial and sensational manner. It is very easy to sensationalize these matters and inflame people's minds. I am attempting to give you a more balanced view, in my opinion, of many of the questions you have raised. But I would not participate in or feed the sensational aspect of it. I am saying that this is a very hard problem that can be dealt with in a reasonable way, and it need not be inflated or identified with Nazism. They need not be confused, because they are really two separate issues.

M.S. Is man in danger of having his freedom of choice and his liberty, alienated by the new psychotropic medicines? Can we take precautions with a democratic society against manipulations of the psyche? How?

J.S. Well, the psyche is manipulated all the time. I was just talking about journalism, for example, or other forms of communication. We are not as free as we think we are. Our psyches are manipulated by rising prices, by inflation, by gasoline shortages, by our educational system—whether it be Church or school—by political parties, by cults, and so on.

M.S. What I mean is that we have more and more sophisticated psychological medicines now. One of my friends in France is now dealing with a new psychotropic drug that's a sexual disinhibitor. And more and more, we are going to have a whole range of drugs that will help control, inhibit, or heighten the realm of human behavior.

J.S. I think the best defense against that will be a healthier and more balanced mind, making it unnecessary to use drugs for these purposes. Of course, those drugs do exist, but they are habit-forming palliatives, mental prostheses. Of course, people feel frustrated about this or that, but the choice exists: if you don't have certain things, you're free to use those that are available. Well, we might consider ourselves fortunate to have the choice to compensate, by means of drugs when necessary. Some people are going to abuse anything. They abuse the health they have; they abuse the food that is available; they abuse the medications that might be used in a balanced and reasonable way. The same thing is true with the use of medication on the part of physicians. Often, there is abuse in lack of administration with care. We have a whole variety of syndromes due to side effects of drug use. I just learned the other day that one of the side effects of chronic use of Dialantin, a drug used for the control of epilepsy, is a recurrence of Lupus erythematosis. I didn't know that. And it's a good medicine. There are hormones for the control of spontaneous abortion that can subject the female fetus to the risk of vaginal cancer as an adolescent. But then we come to the second topic, drugs with a risk of dependence. All these things have been developed as prostheses in a sense, as compensatory devices, just as people use alcohol, tobacco. And alcohol is probably abused much more than drugs. So I tend to come back to our need to understand ourselves and to learn how to use ourselves as instruments, and how to control ourselves.

M.S. Should we not try to put a stop to this burgeoning of psychotropic drugs, drugs that constitute temptation for those who have no will, no self control?

J.S. As long as there is a market for them, they will continue to proliferate, and we will have to examine each one individually. That is why I think it is in everybody's interest to look upon these substances (and those who produce them) not as enemies, but to work out the needs of society, the needs of the individuals in relation to each of these questions. Yes, you are going to have to leave it to somebody. You may need a group of sages to make judgments about the FDA's work.

M.S. Who is going to control the controller, and who will be the guardian? It is endless.

J.S. Exactly. It is endless and therefore, at some point, we may come face to face with the need for a group of people who have everyone's trust and confidence. A council of wise people, a council of elders, if you like.

M.S. It is a concept I discussed once with an American scientist who suggested the creation of a kind of Supreme Court, which would have the power to prevent the enforcement of a law—even enacted—if it were detrimental to man's health and the environment.

J.S. Yes, that's right. That will come.

M.S. I don't think it should be imposed upon the people. In a democratic society, it will come by itself. And it will have to be something in which all of the sectors of society are involved.

M.S. Thus, it will be composed of ecologists, doctors, and so on.

J.S. Anyway, you will need specialists but, above all, you will need people with wisdom and common sense. You see, this is one point at which I invoke wisdom. People who are wise will get the information from specialists and the decisions will have to be solomonic ones; as you say, kind of a Supreme Court, made up of people who can make the best possible judgment with existing knowledge and on the basis of the human values important in a democratic society.

M.S. Can mental illness benefit from manipulations of the psyche as well as from electronic means? Attempts to reprogram?

J.S. I think a more interesting question has to do with what I call mental health, which is distinct from that of mental illness. I see mental health as the maintenance of the state of health, enhancement of health and use of the mind for this purpose. It would be far better obviously to prevent mental illness from developing, to the extent that this would be possible. There are some mental illnesses that are not necessarily linked to life's experiences, to the way in which we are educated, the way in which we are taught to see things. These can be either advantageous or disadvantageous influences on the capacity of the mind to cope. I look at the mind as an adaptive organ, a system for adaptation, in the relationship between the individual and the environment, the individual and the species. In the same way I look upon the immune system as an adaptive system. It can either function appropriately and harmoniously, or inappropriately; it can adapt or fail to adapt. There are some disorders or malfunctions—maladaptive behavior, maladaptive responses—that need to be reprogrammed, if you think the mind is programmed early in life culturally, experientially, leading to some malfunction along the way. One would have to consider forms of treatment appropriate for that type of malfunction. Then, there are others, malfunctions where chemical substances are involved and are necessary for restoring function to a more adaptive type of behavior. There you see the use of Librium and other antidepressants. I see those substances as equivalent to whatever is used for the treatment of gout, diabetes or thyroid disease or some innate metabolic malfunction. Other forms of psychotherapy involve interactions between people. And in my view, all of these are going to be found to be useful in one form or another.

M.S. You do not foresee some great breakthrough in this field?

J.S. No. I say that with vehemence; that is a kind of simplistic idea that I am trying to destroy in every answer I give you. I don't believe in any miracle cure and it triggers a strong response in me; I am allergic to questions relating to simplistic breakthroughs, as if we had no individual responsibility toward health and well-being. As if we must await manna from heaven or the Messiah or some drug or some injection or some simplistic manipulation. You ask me questions that imply that we are being robotized and you ask me whether I see the possibility of our becoming robotized.

M.S. Not necessarily. What I mean is that, at a certain stage of knowledge, we can foresee some advancement because all the elements are there.

J.S. Give me an example and I will show you why the question is inappropriate for this particular set of phenomena.

A Stoic of Our Time

M.S. Antibiotics. That was a discovery by chance. But after all, sulfanamides preceded antibiotics.

J.S. That was also by chance. It is because somebody was studying something else. And the discovery of anti-psychopharmacologic agents took place in the course of looking for antihistamine. So breakthroughs usually occur suddenly and by chance, and by design, because somebody recognized them.

M.S. What about poliomyelitis? We knew the agent.

J.S. And that was a perfectly rational approach.

M.S. There, the breakthrough was the technical manipulation to create a vaccine. But it was a very difficult kind of guessing.

J.S. Not so difficult, not so complicated, at least in my mind.

M.S. What was the complication in this case?

J.S. There was none, as far as I could see, because it was done quite easily. But it did require enough understanding of the nature of the virus, its propagation, its manipulation, to destroy its infectiousness and detect its immunizing capacity. That involved studying the responsiveness of the immune system and properly manipulating it. But there we were dealing with a single, simple exogenous agent. We are dealing here with endogenous factors, i.e., the pattern of responsiveness of the individual as it relates to the individual's genetic and post-genetic determination. That is the same problem with cancer. It arises, to a large extent, out of endogenous programming. The same thing applies to new diseases. All the easy problems have been solved or can be solved; we have no real difficulties with tropical diseases, parasitic infestations. They are more complex, but they are amenable to manipulation and far easier to deal with than the kinds of problems we have just been talking about. However, there are problems linked to controlling parasitic diseases because people are not working on them, as they might, and because of the way people are programmed psychologically, sociologically, politically, and economically. And also because of social, political, economic, and other considerations that have to do with ways of life in some countries. And so, in that sense, we have a problem not unlike the ones we were discussing, mainly the way the mind functions. Because the way the mind functions determines a policy, a response. Therefore, I see a very important link between the kind of problems that exist in the world, the scientific knowledge and technology that either exist or could become available, and the solution to these problems and the administration of the sociopolitical economic factors involved. Therefore, I see a usefulness in the maintenance or the development and maintenance of healthy minds, healthy ways of thinking for the species as well as for the individual. You see the complexity of this problem. And you see immediately that simple technological breakthroughs do not solve problems. We see polio occurring in the developing countries now, in some of them to an extent that exceeds anything we saw in the developed world. Here it is, twenty-eight years after a vaccine was developed. It's fascinating to analyze the reasons for it. I won't go into that now; it would fill a book. Nevertheless, it is from such experiences and such observations and perceptions that I answer you.

M.S. Will morality change because conception and sexuality will be totally dissociated?.

J.S. Is that bad?

M.S. I don't know whether it is bad or not, I have no idea. But it is a fact of life.

J.S. We're not getting into that; it's true but, by the same token, if you go back into the question of the history of sexuality or sexual behavior in different cultures, there is an enormously wide variety of factors. If you consider the Western world (and within it, the Christian world), and if you take each segment and look at it separately, you will recognize that the attitudes toward sexuality in France and in the United States, for instance, are quite different.

M.S. No, not so very different. The problem here is related to the question of whether we are going to build a new ethic on the new perceptions of science. It goes for sexuality, but it also goes for ethical considerations, based on what we already talked about, namely euthanasia. There are quite a lot of problems we have not dealt with in the past, in the pre-scientific world. Will we be able to construct a new morality? We need rules.

J.S. Rules are made after the facts. Rules emerge from facts. Old rules go out and new rules develop. We must, however, retain from the past what is of value for today. I don't believe that these changes take place just because we decide ahead of time we are going to operate under a new set of rules. I believe that the new set of rules arises out of experience and out of necessity. So, we can expect that, because of the possibilities and the capabilities that exist for questions pertaining to sexuality, to the existence of prostheses, grafts and so forth, a whole new set of ethical decisions will emerge and that all the decisions that are made will conform more to the nature of things.

M.S. But there will be laws, there will be a legal basis in democratic countries, and means will have to be found to prepare it. Public opinion does not seem to be ready. What can be done?

J.S. Educate people properly and appropriately and don't feed them these absurd, unrealistic dreams. It is really the new "opiate of the masses." For instance, you ask me whether I can conceive—because of a population explosion, or exhaustion of natural resources—of a habitat for man in space, on the sea. It is one of those questions that I tend to dismiss from serious consideration.

M.S. Your government is investing a couple of million dollars in this.

J.S. Yes, but not necessarily, in my view, to serve as a habitat. I see it more as a way of capturing solar energy in space to transform it into energy that can be used; a way of creating artificial tides, of manufacturing consumer goods, creating marine farms. I would rather ask how we can solve the problems—the population and natural resources problems—right here on earth without this kind of cop-out, this kind of escape. In my view, by developing a sense of responsibility. By living with or dealing with the realities here on earth. So, if we have demographic or other reasons for the exhaustion of natural resources, let us approach that in a rational way.

M.S. I am now speaking to you as a leading immunologist. Is it fantastic to anticipate that immunology—or the immunological type of nonaggressive approach to medicine—will solve some major problems that now require aggressive therapy such as chemotherapy or surgery?

J.S. Immunology will become increasingly important in a variety of branches of medicine. Now, when you mention chemotherapy and surgery, which diseases are you referring to, more particularly?

M.S. Well, cancer.

J.S. There are a whole host of diseases, such as the auto-immune diseases or the immunopathologic disorders, where a very important role is obviously played by the immune system. And it is also true in some endocrinological disorders. As time goes on, more and more will be discovered about the usefulness of immunological knowledge and immunological technology. About advantageously manipulating the immune system, what can be done with chemotherapy, which is becoming more and more sophisticated, and with surgery, which is necessary when things have gone too far. Now, can I imagine the day when immunology could be used to obviate the need for chemotherapy and surgery? I think it is unlikely. I think it is unlikely that the cancer problem will be solved by immunologic means in the same way as the polio problem was solved. It is based upon an entirely different set of causal factors.

M.S. Is computer-controlled public health care a prelude to an even more police-like society of tomorrow, one that would not have any possible escape?

J.S. I do not believe that computer-controlled public health care is a prelude to an even more police-like society. First of all, what do you mean by the computer-controlled health care?

M.S. That everything will be in a computerized file including HLA, for instance, or other features of our genetic code. Then it will be much more accurate than the usual files.

J.S. You know, you can't have it both ways. There are advantages that could be turned into disadvantages, and that's what you are talking about. Computer-controlled public health care is a good thing for prevention and diagnosis or prognosis. Now you bring in something from left field that has nothing to do with the price of beans. This type of fear should be eradicated because I don't believe Nazi-type times will recur. It will largely depend on the type of society, and I think that we are going to resist totalitarianism in every possible way, because nobody wants it. Everyone is aware of the danger. And there will be all kinds of safeguards possible.

M.S. Will the average man be able to exercise biological control over his own body by the use of microprocessors? Is it desirable? I have in mind all kinds of small gadgets that are (or are on the verge of being) sold in drugstores, gadgets controlling different functions of the body.

J.S. Such as?

M.S. Such as, in the case of diabetes, control of the flow of sugar, and the possibility of injecting insulin automatically. There is a whole new field of self-treatment.

J.S. Marvellous!

M.S. You think that's marvellous?

J.S. Every kind of help is marvellous. Are you assuming, are you talking about people doing things for which they do not have the knowledge?

M.S. Of course. It is self-treatment, just as you have self-diagnosis.

J.S. Well, in principle, I am against that.

M.S. You are going to have in every drugstore, automatic devices to take an electrocardiogram, blood pressure, and so on. When I was a practicing doctor, it was not an easy thing to take blood pressure properly.

J.S. Well. First of all, you are talking about one particular situation in one particular country. There are four billion people on the face of the earth. Do you imagine that these things you can buy in a drugstore in France are going to be made available to four billion people on the face of the earth? Of course not. These questions are, in my view, obviously Western-oriented. . . . I just cannot visualize people having ten such gadgets. I do not think there are many people in the Western world who are going to succumb to the fear you are conjuring up here. We must have a more sober approach. I think it's fine that we have, for example, pacemakers. For the same reason, I think it's fine that people can take their own blood pressure and monitor that. So they know if they have hypertension, they can control it as a diabetic can test his or her urine. Now, I don't know anything about the electrocardiograms that people take themselves. It is obvious that they have to be properly interpreted, and I do not know the use to which they will be put. Unless people have arrythmias. And if they have arrythmias, and the electrocardigram monitors the arrythmias, it may be of some use. These continuous electrocardiograms can conceivably be of some value for the physician in that he can make a diagnosis from reading them. I think people exercising control over their own bodies is fine, if they can control their own bodies and their own minds and exercise more self-discipline and self-restraint. What you are talking about is a form of biofeedback, which I see as extremely desirable because it makes people aware of what they are not doing. Or what they are doing that perhaps is disadvantageous.

M.S. One expects so many miracles of the new biology that some, already, are speaking of this discipline as being able to give answers not only to therapeutic problems but also to food, energy, industrial and other needs in tomorrow's world. Do you believe this?

J.S. Well, I think that the new biology has enormous value, but it must be realized it has limits. And it is essential that this be done to coincide with the needs that exist in humankind. Now we have created the needs by the successes that have already been achieved, in human evolution up to this point in time. There are many more people on the face of the earth. We have developed many more devices with which to live: there is housing, urbanization, transportation. And so I can see the new biology as a means toward a better understanding of ourselves, and also for helping us to better cope with life, now and in the future. And it will continue to change. I do not, for a moment, think that life is going to be the same a hundred years from now. Or even twenty or fifty years from now. We are going through such rapid evolution now that we are being challenged. To meet new and growing needs and changing needs, that is where biology is useful, not only for the technical things you have mentioned—therapeutic, food, energy, industrial and other needs—but also to develop a concept of ourselves, our own nature, and our involvement and participation in the evolutionary process. It is basically a philosophy and also a basis for understanding the changing morality that is implicit in the changes that are being brought about.

M.S. Can one really say that we are leaving the era of physics and entering the era of biology?

J.S. Yes. The biological sciences become more important by virtue of the great importance of the physical sciences that preceded them, and I think we are entering an era where the human sciences are going to be even more important. The biological sciences contributing to the human sciences and the physical sciences contributing to the biological sciences: I can see that as a natural evolutionary flow. And that is what I tried to say in my books.

M.S. How do you see the role of the doctor and of "medical power" in tomorrow's society?

J.S. It is going to be different, because the problems and the needs will be different. It will be different in different parts of the world. We are already experiencing serious problems because of the costs that are involved in dealing with disease. We will have to reduce them by improving health. So, if I see the need for physicians to deal with problems of disease, their main task will be to deal with the problems of health, and I mean the enhancement of health.

M.S. Today, public opinion tends to disagree with some aspects of medical power. They want doctors to explain everything. They want to be educated and to discuss everything with the doctor.

J.S. Well, I believe that is a natural evolutionary development, as the public becomes more educated and more informed; more of them go to school, more of them have been exposed to stories in the press, books about the body. They can pick up any encyclopedia, and it tells you much more than it did before. The invention of the printing press. There are more people who are literate. They have radio, television. You'd have to satisfy that need. It is for this reason that I believe people will have to be educated and become partners in the process. They want to become collaborators in the process of health improvement.

Otherwise, we have two antagonistic groups: patients and public, on one hand, and physicians and professionals, on the other. I should say that in a democratic society, any human being deserves the right to know and to be informed. Honestly and truly. And it raises the question as to whether people want to be told that they have cancer, that they are seriously ill. Some want to and some do not. It is a matter that has to be negotiated. And some people prefer knowledge and others prefer illusion. It is just whether you want to think realistically or ideally about these matters. My own preference is to develop understanding, consciousness, awareness, so as to have some influence over our own destiny. We are supposed to have choices. Scientific knowledge and medical knowledge have to become common knowledge. And that is going to take a long time. There is a need for bringing the two groups—doctors and patients—into closer harmony. The consumer and the provider. So that you don't have the consumer lacking in confidence and trust in the provider. The professional should be serving the client, the patient in this case, serving to build and maintain a state of common trust based on explaining things. At least, this is my opinion in the matter, offhand. So the doctor needs to be as human as possible, and as scientific as possible.

M.S. You mean he would be less of a priest and more of a professional.

J.S. No, I said human. You used the word *priest* and I used the word *human*. He should be human and competent. Previously, the doctor was kind of a magician or priest because that was the only methodology that was available. We have a different methodology now, and just because it involves technology does not make him a technician. It should not, because the techniques or technology should be used by the physician; the physician should not be used by the machine. So, it is just a matter of balance, of proportion, of judgment, of wisdom, and the training of the physician should make him increasingly human and competent in the use of the very best methods. The objective is healing, curing, and not doing harm, in the sense of the Hippocratic Oath, the basis for an ethic and morality that is practiced today. The tendency prevailing at the moment—physicians behaving like technicians—has to be countered, to whatever extent possible. One must take into account what the public is going to demand and require. It would be good to be aware of and to anticipate that.

M.S. Can you conceive of a preventive medicine that is not coercive?

J.S. Yes, I can. And that requires education again. If the public is adequately educated and informed, and free of some prejudices (such as the ones prevailing in certain religious groups, who refuse preventive intervention and immunization), I think it is feasible. I do not know by what means, and I would not suggest any in particular. You may have to capture the bishop or high priest of some religious groups and convince them that preventive medicine is God's will if you like, I do not care what device is used, but we must get to it.

M.S. Moses' method on Mt. Sinai. Ethics and hygiene prescribed by God to the Hebrews.

J.S. Exactly. But we have other means of persuasion today. You show them the statistics that can be compared with their own experiences. And that is why poliomyelitis has been eliminated, to a great extent, in those parts of the world where immunization is available. You can see that recent polio outbreaks have occurred within religious groups that have rejected immunization; this happened in the United States, Canada and in some areas of Holland, in some fringes of the protestant Church, some small sects. But that is just an illustration of the ambiguous question of coercion versus noncoercion. Coercion can be gentle and eventually become a new practice.

M.S. Aren't vaccinations, mandatory vaccinations, coercion?

J.S. They may be, but they don't have to be. Vaccination is a personal duty and a duty toward society. It can be fostered by civil education. If not by education, sometimes you have to use mild coercion in the same way that you tell people to drive on the right side of the street, to stop at traffic lights and not to hit people when they cross the street: it's the same kind of coercion. Indeed, you are not free to do anything you like with your automobile. And you have to take the same attitude about whether you are free to get sick and pass on the illness to other people. So, I do not know why, suddenly, the public feels concerned about it, but they do become aware and concerned. And so you attempt to avoid coercion and if people are victimized because they do not like to be coerced they finally discover, by themselves, that epidemics are a problem. It is just a transient period. So my preference would be to use education whenever possible. And where it fails, then you have to do something.

M.S. Persuasion.

J.S. Yes, persuasion: it also performs miracles.

JOSÉ DELGADO
Colorfulness and Exactitude

IT IS SYMBOLIC that one of the most impressive neurologists of our time, the colorful José Delgado, has his laboratories in a hospital complex in the Madrid suburb named after Santiago Ramon y Cajal, the noted Spanish physiologist of the turn of the century who won the Nobel Prize in 1906 for his research on the nervous system.

Built within the decade, "Cajal" is the newest hospital in the Spanish public health system. It has 1700 beds, an ultramodern physical plant and, for Delgado and his international team of researchers, there are 9,000 square feet of space spread over three floors and overlooking the "meseta."

As in the huge American university-cum-medical centers, a research facility grafted onto a complex modern hospital is valuable for its research capabilities and useful in the treatment of disease. Delgado administers a large budget and oversees eight units (Histology, Physiology, Neurochemistry, Bioelectronics, Computers, and Veterinary Medicine) composed of fifty researchers who turn out one hundred publications a year, some of a degree of sophistication—such as those on transdermal brain stimulation—unavailable elsewhere.

"The proximity and exchanges between clinical doctors and research scientists are often responsible for breakthroughs in health care that can be applied rapidly to alleviate human suffering," Delgado emphasizes.

"Properly directed, research can be of great importance in patient care. We facilitate projects in the clinical services and collaborate with clinicians in research in many areas: recording activity of single neurons in patients scheduled for brain surgery; monitoring the general mobility of psychiatric patients to assess the effectiveness of their medication.

"We have many related programs in biochemistry and neurochemistry, and experiments, such as electrical stimulation of the brain of single animals and those in primate colonies, that are equally useful to those in the field. Our electrical engineers are in contact with the clinical services. We develop the instrumentation for therapeutic brain stimulation that can be programmed according to the patient's needs and for telemetered monitoring of many physiological activities, vital for the diagnosis and treatment of the sick."

Delgado is one of the most spectacular scientists of our time. Many people remember—because they were so striking—his experiments on fighting bulls and on monkey colonies in Bermuda.

So I asked him where he was in his amazing, continuing investigation of the animal and human brain, and what was its scope in a global vision of the future.

J.D. I think, looking at the future, that there are three main aspects to our scientific research. Our technology; our working hypotheses; and their medical and philosophical implications.

In order to increase our knowledge of brain functions, we need to improve our technology. Research today is a coordinated effort of many specialists. While naturally we buy much of our research equipment—and import it when necessary—we depend on our bioelectronic engineers not only for maintenance but for development of instrumentation. In a moment I will describe some of our methodology, but first I should give you some background information about the state of brain research.

M.S. You mean how it is conducted in major institutes?

J.D. Yes. While X-rays can give us important information about the brain, and the new scanner machine is vital in locating cerebral

Colorfulness and Exactitude

tumors, to be in direct contact with a behaving neuron we must implant fine wires and cannulae. Implantation of these tiny electrodes—for electrical stimulation—and chemitrodes—for injection of chemical substanc givo us access to any chosen cerebral area. We can stimulate the brain electrically and record spontaneous or evoked activity; we can inject micro amounts of drugs to explore local action and behavioral reactions. Implantation is a simple procedure performed with the animal under general anesthesia. One or two days later brain explorations can begin, and the same animals may be investigated for many years, proving the tolerance of implants and reliability of results.

M.S. But do you always know where your targets are in the brain?

J.D. Implantation of multilead electrodes in an area generally known to be related to certain brain functions is often successful. In other cases, however, a functional exploration of many points of the brain is necessary; for example, to map the location of inhibitory areas which are so important for blocking intractable pain or uncontrollable motor movements. Research in lower animals is basic for later application to human patients with these problems. The best method for step-by-step brain exploration is the implantation of stainless steel tubings, like little chimneys, in holes drilled in the skull. We have placed as many as one hundred of these guides in one animal, making possible the exploration of about three thousand intracerebral points.

M.S. But how is this done?

J.D. A roving electrode is lowered millimeter by millimeter, and brain functions along each tract may be ascertained by electrical stimulation and recording. Each electrode may be fixed at any desired point. With this technology, we have the whole brain in our experimental hands and are able to study it in a very systematic way.

M.S. How does this basic research relate to human behavior?

J.D. These millimeter by millimeter brain-mapping studies are performed with the animals seated in restraining chairs. Naturally their behavioral reactions are limited, and the results of these explorations must be tested in groups of animals free in a colony. To study aggression, dominance, and other types of individual and social behavior which will indicate to a greater extent the functions of specific brain areas, freedom of experimental subjects is necessary. For this purpose we developed wireless, portable radio stimulation units worn by each animal and activated by telemetry.

M.S. And you used these radio stimulators for years in your Gibbon colony in Bermuda? That was spectacular work, to communicate with the brains of unrestrained animals. It is fantastic.

J.D. No, it is physiology. I would like to see animals enjoying conditions as normal as possible. The great problem in brain research in the past has been that it was centered on isolated laboratory animals. Naturally, human beings are not confined to the laboratory. And a full understanding of how our brains work must be based on spontaneous reactions of subjects in a normal environment, or animals in their natural surroundings. Basic laboratory work must be compared with stimulations of the same areas of the brain in the same animals when interacting with their peers in a social colony or when completely free, for example as in our Gibbon colony on the island of Hall in Bermuda.

M.S. Do many scientists use radio stimulators? Are investigations of free groups of animals taking place in other institutes?

J.D. Several miniaturized radio brain stimulators have been proposed, but to our knowledge our system is the best developed and most widely used. Our electronic design includes crystal control of radio signals, FM coding to select channels and parameters of stimulation, and uses an optocoupler permitting electronic separation of stimulating and monitoring circuits. In spite of the instrumental complexity, our unit is only 5 centimeters in diameter by 1.5 centimeters thick, and it monitors all radio brain stimulations.

M.S. Have these units been used in humans? And with what purpose?

J.D. Our miniaturized multichannel radio stimulator and telemetric recorder of electrical activity of the brain called the "stimoceiver" was developed in the 1960s and tested extensively in primates. It was first used for therapy in epileptic patients in 1968. Treatment to inhibit (by stimulation of appropriate areas) or study (by recording the electrical discharges and identifying the problem area) epilepsy can be performed with the patient completely unencumbered. On demand, brain stimulations can be programmed, contingent on the appearance of specific brain activity (such as the onset of an epileptic attack or pain spasm).

M.S. These units have plugs or terminals on top of the head, don't they?

J.D. The original units, yes, but since the 1970s we have been using new technology which is transdermal—under the skin—eliminating any possible inconvenience or infection. It is now possible, with this batteryless, permanently implanted unit—which is a microminiaturized stimoceiver—to be continuously in radio contact with selected areas in the depths of the brain of a free patient.

M.S. How do your units compare with cardiac pacemakers?

J.D. Transdermal stimoceivers are far more complex. A cardiac pacemaker has only one function—to deliver a pulse to maintain the heart beat. Some of these instruments work "on demand," being activated only when the heart beat slows to a certain rate. This procedure is very convenient for long-term therapy because stimulation is controlled by the target organ.

M.S. But the brain is not a simple organ.

J.D. The brain is extremely complex, and while some areas fatigue after only a few seconds of stimulation, others are indefatigable. Some brain structures, therefore, could be electrically controlled for as long as necessary.

M.S. How could this be done—with more machines?

J.D. In the future, minicomputers will be utilized to a greater extent in many fields, including brain stimulation. A decade ago, we demonstrated in a chimpanzee that the brain may be able to inhibit itself through the intermediary of a computer. Brain activity, recognized by a computer, may trigger radio stimulation of a second, inhibitory brain area, which can block the original recorded brain waves.

M.S. What would be the use of this rather spectacular feat?

J.D. Supposing that a computer were programmed to recognize the irregular activity coinciding with the onset of an epileptic attack or other undesirable behavior. It could then trigger excitation of another cerebral point and inhibit—arrest—the unwanted activity. We did this in the chimpanzee by blocking spontaneous electrical waves in a cerebral area called the amygdala. Therefore it can be done, and by making use of the refined equipment already available, we will be able to treat many medical problems in the future, and avoid, for example, unnecessary medication which affects many organs, not only the focus or origin of the specific problem. Patients would no longer have to suffer in anticipation of dreaded attacks; before they were even aware of an incipient crisis, the computer would recognize the warning signals and initiate appropriate treatment.

M.S. That sounds like science fiction, absolutely incredible.

J.D. If our resources were concentrated more substantially on medical research, on helping humanity instead of on designing unnecessary commodities or war machines, we would surpass many of the dreams of science fiction, which are often based on inaccurate or misunderstood data.

M.S. Your ideas about having a computer direct human health or behavior sound rather frightening—how can you explain this to the general public?

J.D. The goal of modern medicine is to treat precisely and exclusively the malfunctioning area and permit the patient a normal and unhindered life. Until now, much medical care has been based on practises such as chronic medication, which may have adverse effects on organs not involved, and surgery, which is the need to eliminate malfunctioning parts of the body that have no known cure. In the future, interference with the whole organism and general trauma should be more effectively avoided. No patient could expect to have doctors and nurses monitoring his heart beat or the electrical activity of his amygdala day and night, yet we have machines that can do this, to ensure immediate, prescribed treatment. This represents the liberation of man from anxiety, unwanted side effects of medicine—a dream for the future, but one that can be realized.

M.S. Computers are very large and very expensive—aren't you imagining a type of medical treatment that would only be available for the privileged few?.

J.D. Whatever is available to the general public depends on how each country decides to spend its resources. That, in turn, depends on how the citizens are educated—whether they would rather have television than sanitation, for example. If technology is given the necessary backing, instrumentation can be produced which is smaller and less expensive and therefore within the possibilities of the masses. Instead of an expensive and bulky computer, a microprocessor could be used. We are now testing a totally implantable model designed to handle sixteen electrode probes for stimulation and/or electrical recording. The program stored in the microprocessor directs the selection of points to be stimulated, contingent on an algorhythmic decision process that detects pre-established electrical patterns in the brain. This technique will permit indefinite duration and total autonomy for the establishment of artificial connections within the brain and for the contingent activation of selected cerebral structures.

M.S. And what about chemical stimulation of the brain, which you mentioned at the beginning of our talk?

J.D. Most of what is known about the chemical activity of our brains is based on studies of the whole brain, injected with some substance, later extracted, and homogenized. Today we are able to inject micro amounts of selected drugs into very restricted neuronal areas, where the drug may be absorbed slowly and its effect on electrical activity or behavior recorded. We can also withdraw liquid from the brain and analyze the chemical composition of different areas during specific types of behavior and under specific conditions. Since chemical and electrical activity are so intertwined, the method of chemitrodes (electrical *and* chemical stimulation) is basic to our understanding of brain function. We can perfuse labelled substances such as tyrosine, and then observe the metabolic activity of each brain structure. The brain chemistry of the awake, acting subject is a whole new, exciting research area of great promise for future therapy.

M.S. What about your working hypotheses? What are the ideas that have directed your research along its present lines?

J.D. One hypothesis is that behavior is not organized in "brain centers," as previously believed, but that our actions are functionally represented by constellations of neurons in different cerebral areas. Until recently, most textbooks referred to centers or areas of the brain, stimulation of which would elicit specific motor movements or types of behavior. More extensive investigations have revealed, however, that there is considerable redundancy in the brain—which often makes possible a recuperation of functions in case their so-called center (one of their constellations of neurons) has been damaged. Multiple foci for specific types of behavior have also been identified, demonstrating that many distant neuronal pools are interconnected and are generally in a state of equilibrium. For example, in the treatment of Parkinson's disease during the last few years, a series of brain areas have been the preferred targets of therapeutic destruction. Evidently they are all involved in triggering the unwanted tremor.

Instead of destruction, I suggest an alternative possibility: establishing and maintaining a bias in the brain—lowering the thermostat, one might say, to avoid overheating, overreaction. Just as sensors may detect the temperature of our bodies, they could detect psychic states, for example. Perhaps we could impose a bias for our psychostats and help control stress and other mental problems.

M.S. But what about treating a specific physical problem?

J.D. Right now I am particularly concerned with the dilemma of pain. If we could trace the pain pathways and identify most of the cerebral areas involved in reception and transmission of painful inputs, if we locate the inhibitory areas, then we will have the information we need about the constellations of areas involved in pain, and more effective therapy will be possible. The blocking of unnecessary pain is a concrete objective of our research. But I think that we could also affect mental states: happiness, sadness, or anxiety. We will be able to control them better once we understand their cerebral bases.

M.S. How can we hope to understand such abstract emotions?

J.D. How can we comprehend any messages? We cannot, without codes. Another hypothesis is that of the existence of cerebral codes of communication. Several recent investigators have claimed that single neurons react differently when certain words are heard—and therefore that there are electrical traces evoked in neuronal activity which could at least in theory be "read" to identify the spoken words that caused them. Apparently there are "information extracting neurons" in the occipital lobe, for example, which respond to edges, movements, and patterns in specific ways. If this is true, why couldn't there be neurons that react specifically to different emotional or psychological states? Almost 20 years ago in our laboratory we found several neurons, out of 100, that fired markedly in a cat's brain when it was confronted with a mouse. These neurons were evidently involved in the cat's emotional reaction. Could we learn to "read" the neuronal discharge and recognized the message "cat excited by presence of mouse"? Any work that brings us closer to a neurophysiological basis of behavior is, to me, very intriguing and worthwhile. At present we are attempting to relate spontaneous behavior and telemetered electrical activity of the brain. First a tremendous amount of work is done, analyzing the behavior of individual monkeys in a colony. Then it is matched with the recorded electrical activity and, with the aid of a computer, the brain activity during certain types of behavior is averaged and statistical significance is estimated. Preliminary results are very interesting, and it appears that we are beginning to unravel this aspect of brain mysteries. At the same time we are telemetering information about general mobility, eye movements, and muscles, and all this data about simultaneous ongoing activities will reveal correlations between specific parts of the electrical activity and specific types of behavior. This work is transcendental if we plan to modify behavior by brain stimulation. We need to know what happens under different circumstances, and where, in the brain.

M.S. Many scientists are so deeply involved in the technical aspects of their work that they are reluctant to speculate about the philosophical or ethical implications. How do you feel about this?

J.D. I believe that it is not only an awesome challenge but an obligation to attempt to evaluate the importance of our efforts and their possible effect on society. For example, all my investigations indicate clearly the interdependence of the elements forming each individual, from single cells to whole organs. And in considering each individual, I realize how completely dependent he is on his companions and environment. It is not a matter of the desirability of cooperation—it is a problem of basic survival. We are not single entities or islands but part of a continuous flow of information from the past to the future. According to this view, we have, if not a theological immortality, certainly a biological immortality. In the Western world, personal imprinting teaches us that we are individually important. Personal destiny and individuality are overemphasized and lead to frustration and aggression. The solution is to understand that we are formed, our minds are formed, mainly by information coming from the outside and this information actually molds the physical basis of our brains.

M.S. In your own field, what do you want to achieve? Many people, in spite of your charm and genius, see you as the father of manipulative, the normative behavorial school. I don't think that your aim is to become the brain behind Big Brother in the future, but what do you really want to achieve in the field of therapeutics and not of conditioned behavior?

J.D. Let me say two things. Number one, you are right: my personality and especially my work have been misinterpreted not only by some scientists but by the mass media. The reason is that they feel the danger of someone who knows too much and therefore could manipulate their own minds. The purpose of my research is exactly the opposite. It is very difficult to get this idea across. If I say that human beings are robots, automatons, what am I trying to do is irritate people, to awaken them so that they can escape from the robotization of human behavior. If you think that you are free, then you will be a robot. If you realize that there are many elements that will determine your own behavior, than you can and should encourage the development of intelligence and individual freedom. First, you must know their determinants. What I am trying to do with my research is to clarify the possibilities and limits of human freedom with the idea that personal freedom and originality are not only self- but socially rewarding, and that therefore we should encourage them for the good of mankind. Freedom is not a physiological property of the brain. On the contrary, behavior is determined by genes and social imprinting. The normal behavior of the brain is to be a slave. If you want to have free people, you must educate them to be free—not to be slaves of society, or even of an ideology, political or religious. If you are a member of the party, you may obey the discipline of your party blindly. This is not freedom. Freedom is only possible when you are taught how to think and act, sometimes even against established modes of society. One of my main purposes has been to investigate the physiological mechanisms of brain and mind in order to apply this knowledge to education. People must be taught to be aware of their emotional mechanisms and thinking processes in order to develop their intelligence and personality, to make their own choices. We must not follow blindly but be aware of the frames of reference inculcated in us by our particular cultures, and try to use them with perspective. I do not pretend to be involved in behavioral engineering—or at least only partially—in the sense that we cannot escape, and this is the principle that people misunderstand. It is biological reality to state that we are subjects of educational engineering. We cannot escape because, in the absence of sensory input, the brain will not be shaped and the mind will not appear. You cannot live if you do not breathe oxygen.

M.S. Now that you have exposed your philosophy as a humanist and not as a brain manipulator, what do you hope to achieve?

J.D. Foremost, a clearer understanding of the human dilemma: our present lack of freedom, and our tremendous potential. It is vital to establish programs of education with a biological basis. The great fallacy today is general acceptance of the idea that babies are born with set personalities. There is, at least in Western societies, too much concern for the individuality of the baby, and at the same time great child abuse, all based on misinformation.

M.S. How would you define a human being?

J.D. When babies are born, they do not talk, walk, recognize their environment, comprehend conversation or signals, plan for the future—none of the so-called signs or qualities of being human have yet developed, and they are completely helpless. A human being needs to communicate, and a baby must learn gradually. It is not born with a personality—tendencies, possibilities, yes, but only through the experience of being well or badly treated will the baby's personality develop. Just as you must feed and care for the baby, selecting everything for him because he cannot choose, you must provide the information to mold his character. Without adequate food, he will die. Without adequate information, his mind will not develop. It is your responsibility. Babies should be treated with much more respect, and tenderness, and consistency. It is our responsibility, that of parents and society. The child cannot escape.

M.S. Then you favor a new morality, a new democratic order?

J.D. Not a new democratic order, a *biological* order. All societies—including capitalistic, socialistic, and third world—must realize the common denominator of all human beings. We all need food, shelter, care, and we need ideas. Basic human reactions can be trained toward cooperation and integration. There will not be one Big Brother—there must be a big brotherhood of all mankind. We share the same biology, the same functions and needs.

M.S. What are your views about genetic manipulation?

J.D. As with atomic energy or any of the other main discoveries in the advancement of civilization, naturally there are risks and possibilities of misuse. There must be safeguards and there may be disasters—explosions, distributions of drugs that cause unsuspected medical problems—but knowledge cannot be suppressed. It will be used by someone, and the best policy is to try and communicate and disperse knowledge among responsible groups that can counterbalance each other and ensure the most careful use.

M.S. Some people say genetic engineering will open a Golden Age, and others see it as the end of the world.

J.D. It could be either, but I hope somewhere in the middle. I hope we can catch up with our technical destructive power by increasing human wisdom and cooperation.

M.S. Do you think it will be possible to live 120 years? Is it something we should wish for?

J.D. I don't think that the span of life is the important problem; the quality of life concerns me more. We should try to live as long as possible provided our lives are personally enjoyable and socially useful.

M.S. In the new biological morality, can euthanasia be part of the picture?

J.D. I think that this is an individual decision, with two aspects. One is to suppress life when you think it is not pleasant, and another is not to prolong life unnecessarily. I believe that we need dignity to live and should be allowed to die with dignity. We certainly need a new conception of what death is. We are taught, and this is cultural imprinting, that death is something horrible, that it is the end, or that we may have the flames of hell waiting for us. I think we should teach our children a new concept of what a human being is, about his biological cycle, about birth and sex and death and all kinds of things which are very natural, from a biological point of view, and that we share with all mankind. If we accept them as part of reality we will not be afraid of being alive or dead. I don't think that we should kill human beings, but I don't think we should prolong life artificially in helpless patients. We need a new kind of education for living and for dying.

M.S. Gerontology is a tremendously important field in medical research, with billions of dollars invested.

J.D. I don't think this is a waste; it is just badly oriented. Gerontology should not have as its purpose the prolongation of life as such, but the re-education of old people to be self-sustaining and useful to society. It is a great mistake to separate this huge number of people from society, typing them as old and useless, without family or social roles. It is cruel, and stupid, because it is wasteful of a great human resource.

M.S. This new biological order, or awareness, seems to be destroying the religious concept, at least in the West—the Christian beliefs. And what about sexuality in the world of tomorrow where you will have contraceptive vaccinations, babies born in test tubes, where conception and sexuality will be completely dissociated?

J.D. You are asking me several questions: one about Judeo-Christian morality and two about sex.

M.S. It is not a question, it is a fact that morality is falling apart with this new biology.

J.D. I don't think so. I think that the mistake is not to re-evaluate the old morality in terms of the new biology. They are not divergent but rather convergent. People who would like to preserve their Judeo-Christian or any other religious morality should adapt and not oppose their beliefs to biological reality. The error has been to put barriers against the advance of science. If you accept the new biological principles and then choose to give them a supernatural destiny, it is up to you. Science is not opposed to religion. Religion is mainly an emotional interpretation of reality inculcated by education.

With respect to your question about sexual behavior, again it is a matter of education, without which we exist at an animalistic, lower level of instinctive behavior. Sex should be considered not as a set of techniques to be learned for self-centered pleasure; even from a hedonistic point of view, for greater pleasure. We should be taught that sex is a means of deeper communication with those we love. We need to give spiritual meaning to this biological function.

M.S. In the world of tomorrow, could you imagine the existence of organ banks and replacement of this or that part of the body, as we change parts of an automobile? We would have a very difficult ethical problem.

J.D. This is not a question of tomorrow but of today. We already have organ banks and there is no problem except that we need more parts, more donors. We need to be more generous and think of ourselves not in the self-centered way as we were educated but as part of the whole. You should derive pleasure from knowing that your own eyes or liver could be used for someone else. I don't see any ethical problem. Organ banks should be expanded and will be excellent for society and for the individual.

M.S. Then we have to educate people to give their bodies as they give their money.

J.D. Exactly.

M.S. If we consider that surgery is barbaric, that chemotherapy is also barbaric—could we imagine a day when medicine will not be aggressive, and will be only immunologic?

J.D. Immunology should expand to solve some body problems, but each branch of medicine will have its own role in the future.

M.S. Do you foresee a world in which people could control their own bodies, and correct deficiencies by themselves?.

J.D. That cannot be answered in general terms. We can learn to control some functions, but medical problems such as tumors or viral disturbances won't be alleviated by the patients. The individual cannot cure himself, because of his lack of knowledge. In medical treatment you need specialization.

M.S. What about preventive medicine? People often refer to it in the context of a police state. Could it be implemented without being coercive?

J.D. You cannot avoid coercion if you are including all regulations by the state. The state must establish rules and laws—you cannot pass a red light, bother your neighbor, rob or kill, and you must pay your taxes—social life is not possible in a complex society unless many basic laws are observed, and this means what some would call coercion and others a lack of freedom. We must differentiate between necessary social regulations, which help preserve individual freedom, and unnecessary, antisocial regulations, which limit freedom of thought or expression in many countries which *are* police states. Laws should protect people and favor their development, not confine or enslave them. What we need is a humanistic civilization in which we can develop our full potential. The old saying, "Know thyself" should be changed to "Construct thyself."

M.S. Are you predicting a kind of mental revolution?

J.D. It is possible, although perhaps utopian, like all hopes for mankind. Thinking for one's self requires a lot of energy and effort, and it is easier to follow the masses and do what one is told. If, however, instead of leaving the future to those in charge of the mass media and having people trained simply to obey commands, so that personality is indeed created by Big Brother, we train people to evaluate information, give them alternatives, give them weapons—techniques—of mental self defense, they *can* become actively involved in the development, the genesis of their own psyches, which I call *psychogenesis*. The discipline of psychogenesis should give the individual awareness of his own mental powers and mechanisms, and therefore greater freedom. Psychogenesis should be taught like geography or physics, but it is even more important because it can help each person to be more individual and to direct his own life.

Psychogenesis is very free, very open. It is not the inculcation of a political system, only the clarification of the biological principles which direct our lives. Once you know them, you can use them in any way that you want. You can control your weight if you want to, and you could do exercises to control your mind. You should be taught how to leave your mind blank, or concentrate deeply. To block out unwanted noise, to control your emotions. You should be taught how to avoid unnecessary mental suffering.

M.S. Doesn't this mean a whole new curriculum?

J.D. No, only the addition of a new subject, which is perhaps the most interesting—a practical, personal application of biology to improve the functioning not only of your body but of your mind—therefore making it possible to educate yourself and free yourself from unnecessary constraints of civilization and the environment. To encourage individual freedom through biological knowledge: this should be a main goal of psychogenesis.

M.S. Are you influenced by oriental philosophies?

J.D. A little, yes, but not completely. The general concept of the individual as part of God or the whole is very meaningful. Integration instead of isolation: for example, right now we are exchanging information and I am part of you and your are a part of me. Probably you are going to remember having been in Madrid. The fact that you have been here is going to modify your own brain slightly. Your neurons are being imprinted, and memories are being established.

M.S. What could be achieved by psychogenesis?

J.D. As in the title of my book, a psychocivilized society in which the maximum of individuality is combined with the maximum of social integration.

HANS KREBS
Weimar's Autumn

THE NAME KREBS is legendary. There isn't a biochemistry or medical student who doesn't know the famous cycle that bears his name. Like Otto Warburg, whose friend and disciple he was, Hans Krebs was one of the great pioneers of today's biochemistry.

Winner of the Nobel Prize for Physiology and Medicine in 1953, he was knighted in England where he took refuge with the advent of Nazism. Until his recent death he directed the laboratory on metabolic research of the Radcliffe Infirmary at Oxford.

He was a man who was curious about everything. He was attentive to everyone, with the rather old-fashioned courtesy that evokes the charm and sweetness of life of another time, before the horrors of the second world war.

For some reason I associate that attitude with what I imagine to be "Weimar" intelligence: Weimar, that autumn of sensitivity and in-

---

Hans Krebs died in 1981.

telligence both German and European, which was perhaps the intellectual highpoint of our time.

M.S. Men have gone to the moon. Why shouldn't we hope for a cure to such serious illnesses as cancer in the near future? In other words, do you believe in a scientific utopia?

H.K. It's very hard to make predictions. For instance, if someone had prophesied only thirty years ago that man would go to the moon, people would have said he was crazy, or in science, for instance, that we could decipher the genetic code, it would have seemed to be absolutely impossible. So, sometimes developments take place that cannot be predicted because the process is gradual. One cannot really foresee in which direction progress will go because much of the progress depends on fundamental discoveries and some depends on simple new ideas that revolutionize our fundamental beliefs, I mean the basic sciences. For instance in medicine, there was a magically new kind of drug in the antibiotics that could not have been predicted until they were actually discovered by making chance observations.

M.S. By chance?

H.K. In a sense, yes. There were indications that some organisms can produce substances that protect them against other substances in the competition for food. They would become ideal drugs in some ways because they affect the specific properties of microorganisms, the bacteria differ from the hosts in very few ways; the basic metabolic processes are the same, but the cellular membrane is different. If penicillin just attacks a certain membrane, that is where the point of attack is. That is why they are ideal: if it is a good drug, it affects the host only in a certain way.

M.S. As a biochemist, what do you foresee in the vast field of cancer, where we know little or nothing of the basic mechanisms?

H.K. Well, important work is going on in many places, but which of this work will be a breakthrough? There is some work being done at Oxford, for instance by Henry Harris, which suggests that there are in cancers specific membrane properties characteristic of cancer, but that is only a preliminary conclusion. My own feeling is that complete control of cancer is out of the question because there are many kinds of cancer. In the control of cancer the area of action is enlightening the public. We know how to control most kinds of lung cancer: stop smoking. But people still smoke and smoke excessively. Although we have this knowledge, persuading people to act on it is a big problem. In Great Britain, although the government propagandizes, it is not very effective; governments want to get the taxes that come with smoking, so on that basis they don't take action. I think it is not a medical or scientific problem in the ordinary sense but one to which sociologists should be making a contribution, in the behavior of human beings. Lung cancer is one thing, but the contamination of the environment, the exposure to dangerous elements is something else. The Americans went too far in forbidding saccharine. There is no evidence that saccharine is a carcinogen when taken in normal doses. I think that it would be more useful to protect children from sugary drinks.

M.S. What seems prospective in the next few years and what is in the realm of utopia?

H.K. By utopia you mean an ideal we can imagine, while prospective is more realistic. A cure for cancer would be utopia. I think the incidence of cancer could be greatly reduced, but it means not merely an effort on the part of the doctors but also of the population as a whole; they have to be educated. But the chance of getting cancer has increased because people live longer and smoke and suffer from exposure to radiation. Cancer has become an old age disease.

M.S. Do you foresee breakthroughs in the future? Not just major scientific ones, from your work and from the work of your colleagues?

H.K. I think there will be breakthroughs in cancer but only by concerted efforts in many quarters. Precisely where the breakthrough will come I think cannot be predicted. I think the main thing is to have good people working on the problem. I'm doing research on cancer, but I'm not concerned much with cancer directly. It was many years ago when I worked with Otto Warburg, in particular on metabolic aspects of malignant tumors. But it turned out that this work was very important for the understanding of certain phenomena yet was not really attacking the essentials of the origins of cancer. What he studied were differences in the metabolism of normal cells and cancer cells. We now know, which wasn't known when he started fifty years ago, that cancer should be looked upon as a chance mutation that changes the control of metabolism and the regulation of cell growth; cancer is an unregulated growth. What we have learned about in more recent years is the control mechanism, and my field of work is essentially the regulation of metabolic processes. It is in that area that the fundamental disorder of cancer is to be looked for. There are so many mutagenic agents and they occur spontaneously. As all living evolution is dependent on mutation, we can hardly expect to control it completely, but we can diminish the risks of mutation. Therefore I believe that a great many cases of cancer can be prevented. If utopia means desirable, of course it is certainly desirable, but I think that it is not practical.

M.S. There are people who want a cure and not prevention because they are too weak to resist the temptations of tobacco or food.

H.K. I think prevention is important. There are several factors for a cure. First, it depends on early diagnosis; then, on using the various treatments, surgery as well as chemotherapy, and I think more chemotherapeutic agents are being developed all the time by industry. I know of cases where chemotherapy has been effective for many years, even though the remission time is usually much less. Leukemia has been controlled quite well. People don't die of the cancer or of the leukemia, they die of the decreased resistance because you cannot specifically inhibit the growth of cancer cells without also inhibiting development of the natural defense mechanism that produces the immune bodies. I think gradual improvement may be possible in chemotherapy, but early diagnosis too depends a great deal on the collaboration of the patient so therefore, enlightening the public is a very important aspect. I think in Britain a fair number of people know now that when they feel a lump in the breast they must see a doctor immediately, but there are still many people who are unaware of the early signs of cancer. From my own experience, I know of several cases where probably much more could have been done had they gone to a doctor earlier. Enlightening the public is a very important matter. There are far too many hypochondriacs who go to a doctor when they're not really ill, but then there are people who neglect the danger signs, cancer in particular because they aren't painful.

M.S. What do you think about the theories of megavitamin therapy of Szent-György and Linus Pauling?

H.K. I don't believe in it. A balanced diet is more important than megavitamin doses. In fact, I had a number of arguments with Pauling personally. I have great admiration for his achievement, and also for his moral courage. He spent a lot of time on peace problems and got a Nobel Prize for peace also, as you know. But we don't agree on this question. Excess vitamin C can be converted to oxalic acid, which can be harmful and produces oxalates and oxalate stones. When you give these excesses of vitamin C the body adapts to the higher level and destroys vitamin C. When you go back to a normal diet, much of the vitamin C is destroyed, it is broken down, it is oxidized, and oxalic acid is one of the products. But it is a relatively minor hazard. I think that on the whole it is fairly harmless, except for the expense. One has to take into account all the psychological factors. If people think they're doing something for themselves, they feel better already.

M.S. Will man of the twenty-first century be more or less aggressive than today?

H.K. I think it all depends on how people are managed by governments. You see, I feel that merely giving some material help doesn't make people less aggressive. People in Western Europe now are much more aggressive than their parents were, although materially they are very much better off. We have youngsters of nineteen or twenty who can afford a car and many other luxuries and who earn ten times as much in actual purchasing power than people did a generation ago, but they still are more restless than ever. These people are materially well off. Not only do they own a lot but they have the security of more or less permanent employment, but they still complain; they are still not satisfied. The trouble is that they don't have what earlier generations did: job satisfaction. They consider the work itself a nuisance instead of deriving job satisfaction from it, and that applies, of course, to many occupations. And I think a great deal can be done in industry to counteract the dehumanizing effect of mechanization and produce job satisfaction. So whether people will become less aggressive will depend upon becoming more happy, and being more happy, among other things, would depend on doing a job that they enjoy, which they find pleasure in. And educating people in the ways to produce happiness: that means that one has to encourage activities of real interest if the job doesn't provide this because it is mechanized. One ought to start very early at the school level and develop hobbies.

H.K. You mean that the big issue tomorrow will not be the job but what to do with spare time?

M.S. Yes, I think that's very important. Especially because the consequences of these policies are military depotism, internal strife, apathy, and an inability to withstand the barbarians' attacks. Finally there's the loss of family and civic values. That is exactly what is happening to us in Europe today, but of course, nations don't learn from history.

M.S. And we also have a problem with the emergence of the third world, and a population explosion there.

H.K. I think the experience with the third world shows that just giving material help is not enough; there must also be some spiritual guidance. The first thing that happens in Africa, is that when they are free, they lose their freedom and become a military dictatorships. The great majority of the new countries of the former French and British areas are no longer democracies; they are corrupt military dictatorships with very few exceptions. And we are still giving enormous material help. In my opinion what we should do is educate the leaders so that they can build up their own industries.

M.S. There are some scientists who believe that with genetic engineering we are entering a Golden Age, and some others think that genetic engineering will bring us to the apocalypse. What are your views on genetic engineering? I don't mean cloning human beings but applied genetics.

H.K. I think genetic engineering can be of great benefit in medical areas, particularly the manufacture of human insulin in unlimited quantities. Insulin does not remove all the consequences of diabetes, especially those concerning the retina and the nervous system, although, some people say that these effects persist because regulation of the insulin supply is abnormal. So it may be that it is not a matter of chemical difference from human insulin but in the regulation of the insulin, the supply.

There are areas of genetic engineering that will be useful; correcting congenital metabolic deficiencies, for example. Cooley's disease, which occurs very frequently in the Far East, is an inborn error in the hemoglobin. Hemoglobin is a protein with two chains: alpha and beta chains. In some cases of Cooley's disease the alpha chain is missing; in other cases the beta chain is missing: there are two types. And as a result the oxygen-binding by hemoglobin is abnormal. It is with the help of the methods of molecular biology that one can synthesize the missing chain in the test tube; one problem is to get it into the red cells. However, to make the missing protein is already a great achievement. I think that one could possibly cure some inborn errors of metabolism.

M.S. And this can be done through genetic engineering?

H.K. Yes. These possibilities seemed to be completely out of range twenty years ago. However, due to the general development of biochemistry and general biology new possibilities have turned up. That's another illustration of the point I made earlier, that it is difficult to make predictions. But I think the phrase you used earlier, a Golden Age, I don't know.

M.S. Some people think that genetic engineering is not only going to affect agriculture but that there will be a new heavy industry with the use of the biomass. Also, in some areas very remote from biology they want to use genetic engineering to separate minerals from ore by biologic means. Some people believe that this is a new era in human mankind, and others are telling us that there is already a backlash, for example, in the "green revolution" in India. And there could be a lot of backlash if we rely too much on genetic engineering. What are your views on the global aspect?

H.K. I think there might be something in it that could produce new industries and make these other industries obsolete. First, a lot of developmental research is necessary, and I think it ought to be explored. Unfortunately, when one talks of genetic engineering, people have in mind that we can affect human beings.

M.S. The cloning of humans.

H.K. Yes, that is silly. It is also very important that human beings are all different. The health of a society depends on individuals not being equal but being different. It is especially important for the youngsters. They waste their spare time. If they are in a herd they will become automatically irresponsible and aggressive. Destroying things is one expression of the emptiness of their lives. The old influences of the churches is gone, the church in the wider sense. People not only used to attend church, but the religious community provided a society to which they belonged and which provided entertainment and a guided society. Now they are in a complete void and don't know what to do, so they get together and become aggressive. Unless something is done about it, they will become more aggressive. I think it has been a mistake to do too little in guiding people's lives in the name of liberty. They are just left alone, and that is not the right concept of liberty.

M.S. You are rather conservative, are you not?

H.K. No, although I think people should have the liberty to choose what they want to do, but they are not given a choice. I'm conservative in some senses, yes, but I think that we should learn from past experiences. I think that adherence to tradition, which is now frowned upon, is something of great importance. The Jews have survived and provided themselves with a kind of continuity of culture—survived all the hazards of being powerless—that has to do with the structure of some tradition of family life, of the Jewish religion, of which I don't know very much. It is mainly a matter of sticking to traditions and referring to the history of the Jews. I believe in the value of maintaining a certain style and tradition, and in that sense I think I am conservative, but that is not in the political sense. I have only recently looked up two books on why the Roman Empire collapsed: one is by Theodore Munson, and the other one is by Edward Gibbon, *The Decline and Fall of the Roman Empire*. They come to the same conclusions: the artificial politics of the Caesars that maintained the appearance of a free republic in which the citizen's only concern was factional rivalry and finally no interest. We could all be killed by the same infection.

M.S. Under the new biological morality, will euthanasia be acceptable, and under what conditions.

H.K. It must be controlled by several doctors and not by one only. You need more than one person to make the decision. You need, of course, the consent of the patient. If there is a case where the suffering is very bad for the patient and for everybody else, it should be possible to shorten life by painless termination. I, of course, would have no objection if that were to be applied to me.

There is the old principle of the sanctity of life and in the very old days there were no criteria except that absolute sanctity. Now we have enough ways and means to take a much broader view and to make decisions as to when life is not worth living.

In Germany about 1920 a psychiatrist, Hochen, discussed whether there are conditions where life should be terminated because it was not worthwhile. Now this idea was grossly misused because Hitler ordered the destruction of huge numbers of people in mental hospitals, and it was a psychiatrist who raised this general principle. It was discussed but not morally discussed. I think it is a valid subject of scientific discussion.

M.S. Shouldn't we be afraid to discuss it in public?

H.K. We should discuss it. The trouble is that politicians very often make use of half-baked ideas; Hitler was the extreme of misuse. I think the law that forbids the use of saccharine in the United States comes from a half-baked understanding. The law was based on the premise that if animals become cancerous because of a substance, that substance must be kept out of food.

M.S. Is it possible or desirable that we should live 120 years?

H.K. I don't think that we can do anything about aging. It is an interesting biological problem, but it is in the nature of life in higher organisms that we age. Only microorganisms can be immortal.

M.S. But still gerontology became a major science. Tens of thousands of scientists are trying to improve not only the quality but the quantity of life.

H.K. That's true and with some success, perhaps, but I think there is a limit and 120 is more than is possible.

M.S. Is it worthwhile? Is it worthwhile to spend billions of dollars and the labor of tens of thousands of scientists to prolong life for a couple of years?

H.K. I think a lot of money is spent on making life more worthwhile. And I think that it is a mistake to think that it would be good for people to retire at sixty or sixty-five. For many people that is acceptable, but many people lose their purpose in life if they have nothing to do. And what's more, a society may not be able to afford to carry so many people who do nothing productive. People are now generally much healthier at the age of seventy than they used to be. Above all, the life expectancy is now over seventy, whereas at the beginning of the century in Western Europe it was only something like thirty-seven. And this figure is due partly to deaths in the first year. The quality of life is also better. But I don't know how much money is spent on gerontology.

H.K. I think it would be worthwhile only if these people find a purpose in life. At the very least that means retaining their jobs much longer than they do now. Retirement is not an ideal way; there must be some way. Also if the society doesn't make them too lonely. One of the worst crimes of Western society compared, for instance, to Asian society, is that here the old people are cut off. The solitude of the old. The attitude is wrong here. In China it is understood; the old are cared for. There is a strong sense of family commitments, family obligations, which is very weak here, in general.

M.S. Will psychotropic drugs limit individual freedom? Can they be used, even in a democratic society, to manipulate the psyche?

M.S. Well, first, there can be no doubt that Valium and Librium are of tremendous benefit, because many individuals who would have to be in mental hospitals can now be socially useful and reasonably happy. I think the "psychotonics" are usually harmless, and the Food and Drug Administration sees to it that what is on the market is not dangerous. I think it is also very important that hashish and the opiates and such drugs be forbidden. That type of drug often gives a temporary euphoria but at the expense of long-term destruction. And about manipulation of the psyche, I know very little about that kind of drug. I can imagine that taking too much coffee can harmful. Americans drink enormous amounts of stimulating drugs of this kind. I don't know of any drugs that make people clever.

M.S. What I call manipulating drugs are drugs that could just transform behavior; behavioral drugs, if you will. As a biochemist, do you think they could exist?

H.K. I would rather look upon this matter as a doctor and I would say that I spent 33 years mainly in medicine and I never lost contact with reality. I think such drugs would be extremely dangerous. And I think that tranquilizers should be under medical control. All drugs should be under strict medical control.

When I was in Germany forty or fifty years ago, aspirin could not be bought without a prescription. Perhaps it is quite harmless and there are other drugs that have no pharmacologically detectable reaction. I think that that is really the crucial point. I wouldn't argue that one should not explore by research. The practical accessibility of drugs to people must be medically controlled.

M.S. Do you foresee psychotropic drugs that will modify behavior?

H.K. In a sense, of course, tranquilizers are behavioral drugs.

M.S. But I mean more specific. For instance, something that elevates the IQ of a child.

H.K. I think that the IQ may be temporarily stimulated, just as many people find that if they drink tea or coffee they can do mental work without falling asleep. I am more in favor of developing mental capacities within natural circumstances. You must given environmental opportunities for people to develop their minds, but drugs are not the way to do this. But you can do a great deal through education. That means opening up fields to the mind and giving people opportunities. This may mean financial assistance for people to study or giving them access to books and other things. This is the way to develop the mind. I think that there is a tremendous amount of scope. But that is just good common sense. Common sense and not science fiction.

M.S. In a world of contraceptive vaccines, will sexuality and reproduction be completely dissociated?

H.K. We can foresee it, yes. It is a highly individual matter. I know of one case, where it turned out the husband was sterile. The husband agreed that his wife should have intercourse with a specific person who was a friend of theirs so the couple could have a baby. But these are all entirely personal, individual arrangements. I think essentially sexuality must remain what it always has been in the evolution of life. It is not the same as the artificial insemination in agriculture where, of course, family life has been replaced by breeding.

M.S. Will prostheses and organ banks allow people to replace parts as they do for automobiles? What ethical considerations would be raised.

H.K. Yes, but up to a point. Prostheses, organ banks exist now. The eye banks are mainly corneal transplants, and that's a very good thing. I think that this is a developmental principle, but there are limits because it is impractical. In heart transplants, the organs could not be kept in a bank; they must come directly from the source. The same is true for livers and for kidneys. These highly specialized organs depend on a constant supply of blood and the constant removal of the products of metabolism. I think the present situation may be somewhat expanded, but I don't think indefinitely, it is not like your analogy of car parts. And I don't think there are any ethical problems.

M.S.   Only practical problems?

H.K.   Yes. An ethical problem I heard a philosopher speak on recently is one in which you kill the person. You see the moral philosophy of utilitarians, the principle of what is good, morally good, is the maximum benefit for the maximum number of people. There is a surgeon who needs two kidneys and one liver for transplants, and there are three people who would benefit. Shouldn't he kill the first healthy person and take away the two kidneys and the liver because it would benefit three people and hurt only one person? Of course it wasn't really typical of a philosophical question, it was far too primitive as he put it, because it is the quality of life and not the number of people. But it would definitely be an ethical problem whether it would be right to risk the life of a person or of many people who voluntarily donate their kidneys while they are alive, if it is a matter of finding a donor for a close relation. Could you remove organs from the dead without advising the surviving family? Here, in Great Britain, they have to have the family's permission. There is a lot of propaganda to make provision while you're still alive and to permit the use of organs. When I say there is no problem, I mean the regulations can easily be worked out, and they will be worked out in this country.

M.S.   Is it possible that man will live in good-sized communities in space or under the sea, either because of diminished natural resources or a population explosion?

H.K.   People realize now that our resources are limited and can easily be exhausted if we do not control the population explosion. I think the policy of preserving resources and limiting the population should be reinforced. What we find is that the best people in the country, the intelligent people, the educated people act in accordance with these principles: they do not set up large families. But the uneducated, the irresponsible, and the criminal produce a large number of children, because they know the state will look after them. And that's happened all over the planet, in Asia, in Africa. Of course, until a generation ago it looked as if the earth's resources were enormous.

M.S.   Will we be able to mine the planets with satellite machines?

H.K.   I don't think so, I don't think we can get significant quantities of any material. We cannot explore the moon or another planet because the conditions are impossible there. We cannot use machinery because we cannot generate electricity or cannot use an internal combustion engine because there is no oxygen, so I think it would be impossible to rely on outside resources. We can explore the idea of the sea and get deeper and deeper, but there will be a limit. The Earth doesn't have the capacity to carry an uncontrolled population.

M.S. How can you explain that the biological revolution that started in the fifties and sixties with molecular biology, discoveries of interferon, the endorphins, so far hasn't had any application in the field of therapeutics?

H.K. To take an example from another field, we know we can get energy by fusion from hydrogen to helium. At the moment we get nuclear energy from power stations only by fission, but in principle we know that it is possible to get energy without getting the radioactivity as a side effect. Now, we can do this on a very small scale, but the physicists are reasonably confident that in twenty or thirty years it may be an industrial proposition. The problems are the simplest kind to formulate: create a very high temperature and maintain it for a certain time. The knowledge that fusion produces a lot of energy goes back to Einstein. But to get this into a working proposition may take at least eighty years, maybe one hundred years. The same is true for molecular biology. Only after tremendous work has been put into it will anything useful turn up, and one shouldn't be disappointed if it doesn't turn up after thirty years or twenty-five years. So in that sense I am an optimist; in spite of great efforts, there hasn't been enough time yet.

M.S. Is computer-controlled health data the prelude to a more police-like state?

H.K. There is no piece of knowledge, no tool that cannot be put to good use and at the same time to bad use. A hammer can be used to kill a person but at the same time the hammer is essential for any workshop. And the same applies to many areas of knowledge and to many tools. I think computing is a tremendously valuable method, but one can also visualize that it can be misused. I know it might occur in particular in America. People have valued their personal information. Stored in computers, it could be recalled by the police or by the government to damage people, to attack them. I see computers in action very frequently. Computing is used in industry and in hospital work and everywhere nowadays. I make use of computers for collecting information. But I think the benefits far outweigh any possible harm. And when people complain that personal information is stored, that is no argument.

M.S. Do you think the new biology could help man not only to conquer disease but also to resolve industrial, energy, and agricultural demands?

H.K. I don't know how much one can expect from the new biology but I think that it is worthwhile investing a lot of money in research to find out more about biological systems. One thing is certain: we can increase food production, especially foods that are scarce, with the use of microorganisms. We can convert starch into a volume of protein by growing moss on solutions of starch and adding ammonia, ammonia we can get from the air and nitric oxygen in unlimited quantities. This is done by a firm in this country, Rank Hovis, and the material is used at the moment for extending meat or making sausages and pies. They can make sausages taste like a steak. It is valuable for humans because it is nourishing, or you could give it to animals. One of my own students was actively involved with the company ICI, converting substances into protein. My friend Melvin Calvin wants to grow trees in the California desert that produce oil. Biology is a new frontier. In the first part of the century, it was the age of physics, with the discovery of radioactivity and nuclear fission. The second part is the period of biology, and you can expect a great deal from it.

M.S. Can you imagine preventive medicine that is not coercive?

H.K. The future of health very much depends upon interdisciplinary cooperation. It is not a matter of medicine only or biology only but of educating the public in looking after themselves better than they did in the past. Educators and sociologists and lawyers should collaborate with biologically and medically trained people beforehand to create new legislation.

NIKO TINBERGEN
Science in a Wrecked World

NIKOLAAS (NIKO) TINBERGEN kept some of the stubborn conviction that fueled his Dutch and Calvinist ancestors, the "Hardheads" of the Reformation, who would ultimately overcome the armies of Charles V and the Inquisition.

In 1973 Nikolaas Tinbergen shared the Nobel Prize for medicine with the two other eminent founders of the then-relatively new experimental discipline ethology—the study of animal behavior—Karl von Frisch and Konrad Lorenz.

The latest joint work of Nikolaas Tinbergen and his wife, concerning autistic children and recalling methods of comparative ethology, marked a radical turning in the treatment of millions of children who had been imprisoned in their solitude.

Like his brother, who is also a Nobel Prize winner, but in economics, Nikolaas Tinbergen was in his own words a "concerned citizen," a

Niko Tinbergen suffered a stroke in 1983 and was therefore unable to review this interview or make final corrections.

citizen of the world whose natural pessimism—born of his patient observation of animal species—was tempered by his hope of seeing man, that most predatory of animals, arrive at the threshold of civilization: cooperation, conviviality, and harmonious existence within the ecosystem.

M.S. Do you believe in the future of science in general, and life sciences in particular? Are we heading for a better world?

N.T. Yes to the first question, no (for the near future) to the last one. I hate having to say this, but I feel that it is no use glossing over my conviction that, while science may help us to think more lucidly and to see more clearly the predicament we, as a species, have created for ourselves, those who govern us will not act on this growing understanding. What seems to me clear is that, as we recognize and begin to tackle the many symptoms of an ecological and social illness that we have brought upon ourselves, the very remedies we apply (however half-heartedly) create more new problems than they solve. I become daily more convinced that our predicament is getting worse, literally from day to day.

M.S. Which we cannot master any more?

N.T. That's what I think, although I have not entirely given up hope. But I cannot understand why, for instance, Barbara Ward* and Peter Medawar** are so optimistic.

M.S. Even in the area of health, i.e., in its extensive application?

N.T. Here too I am not optimistic. For instance, it seems in practice impossible to make the medical world more interested in prevention, nor do most doctors have any idea about the harmful effects on health of our deteriorating environment. But I would rather limit myself today to matters on which I feel more competent to pronounce.

M.S. Let us talk about twenty-first century man as he should be shaped, by better education and in a better environment.

N.T. Without claiming to say anything new, I feel it is necessary to keep stressing that many manmade changes in our physical and social environment have innumerable deleterious side effects. These are now becoming so harmful that I doubt whether even our adaptable species can remain healthy, or even survive, under the new conditions. We have created too many and too severe physical and social pressures. I believe that we know enough about "human nature" to have good reason to fear that, as the new stresses grow even further, people

*Progress for a Small Planet (London: Penguin, 1979).
** The Future of Man (New York: Basic Books, 1959).

will resort to the use of force and violence to ensure tolerable living conditions for themselves. In such a climate, social and political ideologies become very powerful. As a result, we all live in fear and distrust of others.

M.S. Nothing new so far. That's been man's behavior in the past as well.

N.T. I agree, but the dangers are now several orders of magnitude greater. We will be motivated, are already motivated, by ideologies which might or might not become more fanatic than the religious motivations of the past, but whereas much violence in the past was due to mere greed, more people nowadays, especially in Third World countries, will want to fight to obtain the minimum necessities of life. And when you are near death, you may overcome your finest altruistic scruples and steal from your best and equally desperately situated friends. I also believe that we would already have had World War III if it were not for the possession of nuclear arms. I do not share the view that mutual threatening and saber rattling will prevent war, but I derive some hope from the knowledge that for the first time in history the superpowers are afraid of their *own* weapons. In an all-out nuclear war, the deadly radiation would travel around the globe in no time and kill or mutilate friend as well as foe. Excluding liberation wars, which may have to be conducted for some time yet, we must hope that people will come to see more and more clearly that nowdays war does not pay, since it will be deadly for every participant.

But even if we can avoid war, we will have to make very drastic changes in our life-styles if we are not to succumb to overexploitation of vital resources and to pollution, the latter again physical and mental. And the most urgent perhaps of these changes will be the stabilization, or preferably reduction, of populations.

M.S. As a scientist, you don't seem to grant great credit to possible resources of science. Don't you think we could find ways of survival by and through science?

N.T. I agree with those who say that science has the potential, in many cases already the actual ability, to overcome these difficulties. What is lacking is the will and the foresight. And the will is not there because people do not see that the necessary short-term remedies, which would require great sacrifices, would be in their own long-term interest.

M.S. What is your plan for the world of the future?

N.T. I can't claim to have a comprehensive proposal; no one has so far produced one, even though Goldsmith's "Blueprint for Survival"* and later, amended sketches are clearly moves in the right direction. What is clear, of course, is that a radical reshaping of our society will be necessary, based on a revolutionary change in our values. The main difficulty is perhaps that once we are used to a luxury, we refuse to do without it. Also, it is one thing for us, industrialized nations, to harm ourselves, but we have managed to brainwash most Third World countries into wanting to imitate us, with our mistakes.

M.S. What you are saying seems, however, to be too extreme. The world population seems to be stabilizing, and there are some hopes as to the finding of substitutes for fossil energy sources and for nonrenewable natural resources of the planet, thanks to creative scientific efforts.

N.T. With respect, I have to disagree. The world's population is *not* stabilizing at the moment, it is only in the affluent countries that we see a flattening out of the population curve, and even this may be temporary. In Africa, Asia, South America, numbers are still growing fast. And we need not disagree about the seriousness of the scarcity and hence the cost of energy, nor about the terrible dangers of pollution. We know, for instance, that our refuse dumps all over Britain have deadly poisons in them that will trickle through into our drinking water; we *know* that no safe way of disposing of our nuclear wastes has yet been devised. Yet we continue poisoning ourselves and our children and grandchildren. Science may know some answers, but man is still going foolishly for short-term profit.

M.S. Individually, we can foresee the medium and long term.

N.T. That's what's so strange. As individuals we all take out insurance policies against the mere *possibility* that we *may* fall ill or die, paying quite heavy premiums. And foresters are used to investing in massive planting programs that won't yield anything until thirty to one hundred years later. Yet we are not willing to tighten our belts now for the prevention, or at least amelioration, of the effects of our present "rape of the earth" or of our accumulating pollution.

M.S. I had never thought of viewing the problem in insurance terms.

N.T. I think this analogy would be of some use to politicans who have to "sell" short-term sacrifices to their voters and still get elected. The issues at stake are not at all difficult to understand, so that, given the will, politicians could gain votes if only they explained the basic issues clearly.

*Goldsmith, E. et al. "A Blueprint for Survival," in *The Economist* (2, 2–42, 1972).

M.S. And if, to mention only one example, we have no other choice but to "go nuclear," as a means of producing energy?

N.T. We may well *have* a choice, to begin with. We must learn to live on a lower level of affluence. I think this is imperative. The psychological obstacles are difficult to overcome, though. Few of us would be prepared to return to the modest living standard of, say, fifty years ago (when we were not at all unhappy), let alone to that of more "primitive," yet materially well provided for and in many respects happy people. Now that we have our color TV and our cars, we won't voluntarily give them up. They have become "necessities of life." Also, the livelihood of millions is now dependent on the continuation of the manufacturing of these luxury goods. The difficulties involved in turning back or in designing a policy for qualitative growth—better and safer cars, control of deadly pollution, healthier food—instead of merely quantitative growth—more cars, more international tourism, faster planes—are very great. Yet we will have to reorient our technological efforts. And we have the technical potential to do that.

M.S. Is it not utopian to imagine a radical change possible in our society without coercion?

N.T. Yes, it may well be, and that is why I am so pessimistic. We are faced with an extremely difficult dilemma. One the one hand, democracies act too slowly, and are powerless as long as the voters are not educated in these issues. On the other hand if we want to ensure happier conditions, we may have to have more autocratic governments. But they turn into dictatorships, and though these might begin by being "benevolent" and wise," they have so far always deteriorated into tyrannies, which only revolutions or wars can overthrow. Even in our present democracies, governments and "states within states" such as, say, the armament industries, the military, and the nuclear power station builders manage to do things that educated populations would not tolerate.

M.S. So-called socialist societies, such as the USSR, are dictatorships that cannot make use of their power to build a new man, a new society. There we see the caricature of our own society, with the same rush for a private car, the same pollution and environmental destruction, even worse. So who is working on the model for a decent and viable society for tomorrow?

N.T. Not, as far as I can see, the so-called socialist Soviet Union. But interesting and promising attempts at designs for ecologically sound, no-growth societies have already been made—think of the "Blue print for Survival," the "Meadows Plan"* and others—but there are many scientists who think that these plans are no more than first fumbling attempts. Yet I am convinced that scientists can design healthier courses for humanity than the potentially lethal one we are following now. The difficulty is to "sell" them to the voters. Even if we had politicians who understood our predicament (a big "if"), it would be political suicide for them to base their platforms on long-term changes of course. And we must not forget, of course, that even if a saner policy could be introduced *within* any nation, there is still the anarchy between nations, which makes such attempts risky, for they would have to involve reduction of armaments and a deflection of the massive defense budgets to productive work. Even so, I think we could learn a great deal from what China is doing now. Or, rather, was doing until recently, when it too began to produce nuclear arms.

For those who see the "human predicament" for what it is, and who know that, *unless we change our ways,* we are doomed (which is radically different from saying that we *are* doomed), there is another important dilemma. On the one hand we have the duty to spread information about the lethal aspects of our cultural evolution (for unless people are scared out of their wits they won't want to change their life-style); on the other hand we must avoid paralyzing people with a sense of doom. This will require extremely difficult tightrope walking. Barbara Ward's new book, *Progress for a Small Planet,* is such an attempt. Yet I doubt whether it will be more than a little step toward the ultimate goal.

M.S. You think that in man's subconscious, superimposed on the idea of individual death is that of the whole of mankind?

N.T. Yes, I know that many young people live with that possibility in mind. They reckon with the likelihood that their children will suffer from, say, pollution effects and may well die of them; they know full well that there is a great chance of a nuclear war. Yet they can live with the idea of *death,* even the death of their children. What is much worse is to live with surviving but terribly crippled children, mentally or physically. Many survivors of a famine are damaged for life. When you realize that before the coming of modern medicine most people had to accept that the majority of their children would not survive, you can understand that we are better adapted to losing our children than to living with irreparably ill children. It is partly because of this new awareness that quite a number of young couples decide not to have children at all.

*Meadows, D.H., et al. *The Limits to Growth* (London: Potomac Associates, 1972).

M.S. You belong to the school of the most pessimistic futurologists. The optimistic ones believe that mankind will find a way out of this mess. If we are demographically too numerous, we shall colonize oceans and planets, neutralize nuclear danger, and find new raw materials and new nutritional sources, thanks to biology and chemistry.

N.T. You find this kind of irresponsible optimism only among the uninformed and among those scientists who do not look beyond their own, often narrow field of competence. Anyone who reads what the admirable World Watch Institute* records about the many aspects of population, depletion of resources, and pollution, will know that there is only one justified conclusion: unless we change our ways drastically and very soon, disaster on an unprecedented scale will happen. One can be lightheartedly sceptical about forecasts, but one has to distinguish between a certain prediction of the kind "Tomorrow the sun will rise at such and such time" and *"If we do not change our ways, then* we will deplete our fuels, our food will continue to become unhealthier, our population will grow disastrously larger etc." The prediction has to be taken seriously because so far humanity has not shown any inclination to change its ways.

M.S. If you go to Japan, you can see how Western countries will live someday: lack of space, no healthy air, but a lot of gadgets which make daily life easier.

N.T. I agree, although we must not forget that Japanese and other peoples, mainly in Asia, can stand population densities which we in the West find unbearable.

N.T. I don't know. That may be a cultural phenomenon, a thing you can learn up to a point.

M.S. What about the Dutch, who have an almost Asian demographic density? What about populations living in huge cities of Europe and North America?

N.T. I agree that crowding as such would not be intolerable, but then I am not convinced that the modern city dweller is very happy. But returning to pollution, we must not forget that, for instance, the Japanese began to build their most polluting chemical plants on the continent of Asia, and we in Britain are foolish enough to import Japanese nuclear waste.

M.S. By assuming that the Western world finds scientific solutions for its survival—wise solutions such as demographic control, clever use of natural resources, thereby creating a more harmonious and egal-

*World Watch papers and books are available from the World Watch Institute, 1776 Massachusetts Avenue, N.W., Washington, D.C. 20036.

Science in a Wrecked World 269

itarian society—while the rest of the world turns into chaos, are we not going to have some kind of war between races on a worldwide scale?

N.T. A major and, may I say, unwarranted assumption. But yes, war does look very likely. As I said, we dare not start unilateral disarmament (only small countries can do that, and only as long as the big powers are more or less in balance). A disarmed Western Europe, for instance, would either be swallowed up by Russia or by the United States, or become the battleground between Russia and the United States. And these two don't dream of slowing down their arms production—neither do we, for that matter. In 1976 I was told, on good authority, that each of the two superpowers was producing *three* new nuclear warheads each day!

But it looks as if, at the moment, we have no choice politically whether or not Europe disarms; the two superpowers don't because neither trusts the other. Even so, there is a ray of hope in the fact that the leaders of America and Russia seem to know that they have to fear not only their opponent's power, but the lethal effects of their *own* arms, even if they were to explode on the other's territory. As I said, radiation travels around the world in no time. This is the deterrent effect of the moment. But its efficacy depends on the leaders' keeping a cool and sane head. We have been told that during the last weeks of Nixon's presidency he was little by little deprived of access to "the nuclear button." But Hitler might have sent us all to our deaths if he had had nuclear weapons.

M.S. Well, in your new ecological organization, Europe would not produce nuclear weapons, not even nuclear energy for peaceful use.

N.T. I do think it folly for Europe to manufacture its own nuclear "deterrent," and even of very doubtful wisdom to have American rockets based here. Also it is becoming more clear every day that the power stations for "peaceful" use of nuclear energy have been built with unseemly and irresponsible haste. Those who defend the taking of "reasonable risks" forget that what is reasonable for one single kind of pollution becomes very dangerous when it coincides, as it does here, with so many other kinds of pollution. The combined effect of, say, chemical effluents, food additives, deficiencies in many processed foods, the massive use of drugs and antibiotics, car exhaust and other sources of heavy metals, and so on and so on, is bad enough, but in addition there is the synergic action of these agents. Our population is already far from fit, and easy prey to all kind of diseases and parasites. The whole basic attitude is wrong that shrugs off the many warnings by arguing that "anything must be accepted until and unless it is shown to be harmful." New processes should *not* be mass applied until they have been tested and found to do no or little harm. There is a fundamental difference between these two attitudes. Since the earth as a whole is involved, this experimental plot attitude of "let's see what happens if we do this or that" is now outdated; it is foolish to treat the *entire world* as an experiment.

M.S. Then, why nuclear weapons only in American hands? So that they protect us, and we build the best possible world under the shield of the American nuclear umbrella, without our having to bear the inconveniences and costs?

N.T. I know that we are confronted with impossible alternatives, as long as we have not set up universal world order. Yet I remain convinced of the danger of a nuclear war. Even if nuclear weapons do not physically eliminate world populations, all survivors will be genetically damaged, which would lead first to unimaginable misery, then to extinction.

M.S. Could not these survivors be adapted artificially? This leads us to review the possibilities and the questions raised by genetic engineering.

N.T. I do not believe in its constructive power. I think our adaptability is one of such complexity that playing around with our genetic "endowment" is likely to provoke irreparable disaster. Apart from this, who should have the right to decide on what is a "better" genome for our species? Nobody must be allowed to play God. In my view, the applications of genetic engineering will be restricted to very limited, isolated tasks, such as rendering a parasite harmless, or perhaps constructing better antibodies against certain diseases. That is a far cry from breeding a genetically new and viable human race!

M.S. Let us forget about cloning and all those more or less eugenic science fiction stories on the new man. A genetic genius may help us to fight illnesses, to increase our food and raw material resources by increasing the biomass. Isn't that a promising avenue that scientific genius should explore?

N.T. I think that such limited application may well become even more important than it is already, but let us not try to create a supposedly better world by genetically changing man, while we have not even tried to adapt him to his present environment, or rather our environment to our real needs. Let us start by controlling populations, by sparing our resources, and by reducing pollution. Much hope has been placed in the "green revolution," in Asia, for instance. It is now recognized that its undeniable initial success may well have been a flash in the pan because the new strains face long-term dangers that it may not be possible to meet.

In the Middle West, the many fine strains of hybrid corn obtained through genetical manipulations are said to be deficient in some vitamins—those which are normally found in traditional kinds. I could cite many examples. Every new technology creates some advantages but often has disadvantages, which in their turn have to be corrected by still another technology. That is an endless, proliferating set of ever more chain reactions. The conquest of malaria leads to a population explosion; the use of antibiotics to the appearance of resistant strains of parasites, and so on.

M.S. Can we not find a balance between risks and advantages?

N.T. If we can measure both, yes. But even so, one should ask how one can measure happiness and balance it against, say, more and faster planes. I do believe that science has the potential to become a tool for improving man's lot; it is the wisdom we need for deciding how to apply this new tool that is still lacking. For the moment we are proceeding in a very shortsighted and completely reckless way. We have to reconsider our entire system of values. And what is even more worrying is that so many Third World countries do their utmost to mimic our ways, which by now should have been recognized as nonviable.

M.S. With regard to political and economic anarchy, couldn't the scientific community constitute a counter-power? Let there be ten, a hundred Pugwash conferences.

N.T. Pugwash is just one swallow which does not make a summer. Scientists have had and still have no impact, whatsoever, on political power. On the whole they are crying in the desert, or they fail to consider the social consequences of their work. Many of them begin to wonder whether it is worth saying or doing anything. Whether it would not be better to let the world sink to such a hopeless situation that they will finally discover how deep the abyss is they have created. The scientific world has never been faced with such a critical situation, except, maybe, when the atom bomb was invented. I believe that men like Einstein and Oppenheimer suffered the worst tortures. Just remember Einstein's letter to Roosevelt, wherein he proposed the atom bomb to the American political and military establishment. We know what terrible step Einstein had to take. A fateful decision. Oppenheimer was tormented by it to his death.

M.S. If I understand you correctly, suppose you had been in Einstein's position, would you not have suggested using the bomb or would you have refused to make it? Remember that the Nazis were just about to master the atom. This would have meant that Nazism would have conquered the whole world, exterminated entire populations—tyranny for a very long time.

N.T. I don't know. It was a terrible choice. I think I would have done as Einstein did. I know only too well that the alternative was slavery of the worst kind. But don't forget: Hitler had no atomic weapons. Now we live in an entirely different situation.

M.S. Forgive me for finding your position rather illogical. If the source of absolute evil is the atom bomb and not the power using this weapon, world destiny would have been worse—if the atom bomb had been in Nazi hands. Moreover, we would have had the same problems today. Anyway, the evil genie was out of the bottle and there was no way to make him return.

N.T. I don't see that I am *illogical*. It is true that I see no way out. What I am certain about is that an all-out nuclear war now would be the end of mankind. There might be survivors, but as I said, they would certainly be genetically damaged. And they would not even get the chance to reproduce, for no one can survive sanely in shelters, as civil defense enthusiasts seem to think.

M.S. We believe that, as long as there is no nuclear dissemination, the higher authorities of both blocs will be wise enough not to press the button.

N.T. But for how long can we avoid or prevent such dissemination? For the moment it would seem to be impossible. I repeat: unless we change our way of life, including international disorder, I see no future for mankind. It looks very much as if we are an evolutionary freak, a mistake that can live only for a very short time—the human race has existed no longer than a couple of millions of years, a geological second.

M.S. We would already be dinosaurs.

N.T. I feel that the realization of the near-certainty that our grandchildren will have to endure terrible suffering is very difficult to bear.

M.S. Why such anticipation then? What will be, will be.

N.T. Well, that is more easily said than done. And we return to the dilemma: What is our duty? Tell them what we think we know and make them more unhappy. Yet without fear of a possibly unbearable future, people won't change their ways.

M.S. Before such an apocalyptic view, everything becomes absurd. And especially think of the thousands of scientists who do research on gerontology, to extend the lives of old people.

N.T. It is one thing to extend people's life span, quite another to prolong the suffering of the infirm. The ideal of the medical world to extend the duration of individual lives without being able to keep old people sane and healthy is already a source of agony for many old people. I know. As long as your mind remains reasonably clear and you can still, say, dig your vegetable garden, it could be quite acceptable to grow very old; wise old people might even be of great use to the community, however little they can do to earn their own bread. But as things are developing now, the growing proportion in our population of ailing and unhappy people presents a very considerable ethical problem.

M.S. The aim of gerontology today appears to be less prolongation than betterment of life quality.

N.T. Unless gerontology is based on better preventive medicine than we have at present, it cannot be of much help, since it takes care of people who are already sick. I am convinced that most people in the Western countries are unhealthy, at present. My friend Konrad Lorenz told me a significant story. He spent four years as a prisoner of war in the USSR, as an army doctor. He saw some of his Russian colleagues refuse to amputate gangrenous legs of German soldiers. He thought at first that this was mere callousness, but he soon noticed that the Russian wounded were treated in the same way. More of the Russians recovered than of the Germans; they were on average simply tougher, healthier than the Germans, but that was a fact the Russian doctors had no way of knowing then. This tallies with what an American visitor said after a long stay in Russia: "I am no longer afraid of the great numbers of Russians, nor of their technology, but of their health." In case of a conflict, the endurance and health of the Russian soldier might well be decisive. From a health point of view Western man—Europeans, Americans—are probably fairly decadent. Ill health (and lack of fighting motivation) lose Vietnam-type wars.

M.S. Because life is too easy, they eat too much . . .

N.T. Mainly for those reasons, but probably from an enormous complexity of reasons: too rich and poorly adapted food, the car, a bad way of living, and perhaps also the fact that we allow the people who are poorly gifted genetically to procreate.

M.S. Is that eugenics?

N.T. Positive eugenics—the deliberate selection of "quality" people for breeding purposes is of course ethically and practically simply out of the question and should be rejected in principle. Negative eugenics—say the voluntary sterilization of bearers of known serious genetic defects—might be a possibility, but even that issue is fraught with difficulties. What I was speaking of is *not* the genetic qualities of people, but their reduced fitness due to the unhealthy living conditions of over-affluent society. Phenotypic, not genetic changes.

M.S. I am not as certain as you are as to the Russians' good health in general. Indeed, there are some disturbing opinions from dissident doctors who have sought refuge in Western countries, but let's admit it. By the way, how can we know what is the "general" health condition of a population? What are the criteria? What can make you say that Western populations are physically decadent? The fact that, in sports exhibitions, delegations of the Eastern Bloc win a lot of medals is neither significant nor convincing as to the general health condition, especially when one knows the "breeding" methods used in those countries for the sportsmen.

N.T. Of course Olympic medals do not reflect merely health. But I do maintain that we fool ourselves when we think that we, members of our over-affluent societies, are tough merely *because* we are kept alive longer.

M.S. This being so, we already have a kind of unmentioned euthanasia when, in hospitals, they let seriously ill people die by depriving them of intensive care.

N.T. I am sure you are right. This decision is often made when there is no brain activity anymore, although other body functions remain. But I'm thinking mainly of the form of active euthanasia where people are killed, or allowed to die, after having requested it. I think even this is extremely dangerous. What one has decided upon, one day, fully aware and lucid, might not be true on another day. Moreover, by asking people to accept euthanasia, you are influencing them, you may make them wish to die. As a result many people might be scared to go to a hospital or to consult a doctor, just because they will not trust the medical institutions any more. This issue has many ramifications.

M.S. What about those who request it, in full consciousness? I am now thinking of some of my Californian friends, in some scientific circles, who have signed a request for euthanasia, through a deed executed and authenticated by a notary. Those friends carry this document, which gives permission to a doctor to terminate their life under certain circumstances

N.T. This does not solve the problem for me. The moral problem

Science in a Wrecked World

remains, certainly for the doctor. In itself, science is a "good thing," but a society which is too much impregnated and controlled by science alone becomes a society of robots, with distorted values or even without values. The problem of euthanasia is a question for individual and social ethics, and scientists are no more qualified than any other citizens to pronounce on it.

M.S. They are, however, trying to standardize and moralize something which is unavoidable.

N.T. Obviously, things are not black and white. If the power to decide is solely in the hands of doctors and nurses, one trusts a group of people who are on the whole competent, decent, sensitive and humane. This will not, however, eliminate abuses completely. If you admit that, as a consequence, many people become deathly frightened of hospitals and doctors, then ideal euthanasia, done with the best care and caution, will be a questionable one anyway. Gentle death, thanks to medical euthanasia, is no guarantee to me that distrust will not prevent people from seeking normal medical help.

M.S. You do not trust medicine or doctors.

N.T. Frankly, no, not too much. I'll try to condense in a few words what is in reality a very complex argument. First of all, I think that doctors have too much power because, in this secular world, they are trusted by many people as priests as well as healers, as health technicians. Then, too much emphasis is placed on curing the sick, and far too little on prevention of illness. Then, the peculiar position of doctors in our society, and their training as well, combine to make them assume an overbearing and/or condescending manner, which inevitably leads to an arrogant attitude. Of course, many patients accept and even want this. Further, the education of physicians is in my view seriously outdated, and priorities are wrong. There is a growing admiration for the status of specialist and for research, and a diminishing respect for the general practitioner. Yet the university training of doctors, quite unlike that of other scientists, does not really prepare them for research. Also, the growing specialization leads to a very disturbing narrowness of mind; each specialist seems to be interested only in his own field and often overlooks quite conspicuous symptoms, or ignores them, or squeezes them into his little picture of illness. This growing specialization goes hand in hand with increasing reliance on tests, and an alarming loss, in the newer generation of doctors, of "clinical perceptiveness," i.e., of the ability to assess by "mere" practiced looking at a person, whether anything might be wrong with his health. Finally I should mention the extreme conservatism of the medical profession, which causes a delay of decades in the acceptance of new views.

M.S. You don't like doctors, do you?

N.T. I like and respect many doctors I know. But the profession as a whole does not strike me as very admirable. For one thing the good general practitioner tends to disappear or to become grossly overloaded. In addition, many of them have been brainwashed into a ridiculous respect for the specialist. I have seen the totally unjustified humility, the blind acceptance by middle-aged, experienced GPs when faced with what a whippersnapper of a young, still learning, and already very narrow specialist decrees. There is a lot that's very wrong in the field of medicine.

M.S. As is true for most professions.

N.T. Very much more true for doctors. I do not judge individuals, but I do notice a crisis, a critical situation which makes me speak of a state of malaise in medicine.

M.S. Do you anticipate better medicine in the future?

N.T. One could spend one's life trying to answer this question. I should like to state briefly that I would appreciate it if medical resources were aimed more at prevention than at treatment of diseases. This is no easy task, all the more because most people think (doctors included) that, after all, we are a population in good health. I have even heard rather silly statements by "politicians" such as, "The British are in very good health, because they have so many doctors."

M.S. The words "preventive medicine" have lost some of their value. One thinks of an extra bureaucracy, an inefficient one, which will be added up to the one we already have.

N.T. I think that medical prejudice and even jealousy have much to do with this rejection. The administrative organization of medicine—curative or preventive—is one thing; refusal of prevention because it disturbs some habits is another. I shall give one example. There is a vast and rapidly growing body of literature on nutritional research. This, unfortunately, is largely ignored by the medical profession. I have, for instance, heard senior doctors say, in a kind of condescending tone: "You know, in modern society vitamin deficiencies have disappeared." Yet it is evident that, on the contrary, there is a resurgence of such deficiencies. This is largely due to the way in which modern "convenience foods" are processed, in short to make things easy and profitable for the supermarkets. In fairness to the doctors, one must admit that it is difficult for them to spot the real advances from the many quack medicines they hear about in their voluminous junk mail. Also most doctors simply have no time to read about new developments; either because they are GPs with a large practice or specialists who want to make money. These and many other circumstances make the switch to better prevention difficult to achieve. I have met with some of these attitudes in our own work.

M.S. Do you allude to your present work on childhood autism?

N.T. Among other things. My wife and I, after many years of research, have written a book on our work on this very peculiar disease of childhood. It has taken us quite long, and it is now at last done. I will not go into details, but we have found the most astonishing signs of conservatism, narrow-mindedness and arrogance. But my criticism applies to at least some other areas of medicine as well.

M.S. Does health begin with children's education?

N.T. I propose that we omit this—it is far too complex an issue to be dealt with in a few words.

M.S. Do you think that the family as it is now has a future?

N.T. It has to have one or society will become even sicker than it is. We will not have a sound society without a sound family.

M.S. Are you thinking of the monogamous family, coming from the Judeo-Christian tradition? Many already think that it is condemned to the benefit of sexual communities, generalized forms of promiscuity with children's education by the collectivity . . .

N.T. Recent work on child development begins to show more and more clearly that children develop best when they grow up in a family—whether monogamous or bigamous. Nothing to do with the Judeo-Christian tradition; the system is much, much older. But by "family" I mean the "extended family," i.e., father, mother and children, but also grandparents and on the periphery aunts and uncles. Where children develop best is in a group of extended families who live close together and know each other intimately.

M.S. Is it the Christian or the scientist who speaks now?

N.T. The point I want to make is that, for mental health, the "virtual break-up of the family" (as the American educationalist Uri Bronfenbrenner has called it) which is now happening in our most advanced anonymous societies (the U.S.A., as always, in the lead, both with good and with bad developments of the modern industrial society) seems to me a very dangerous development because it damages children emotionally, including many of those who within the near future will run society. What we feel we are discovering in our studies of autism and related disorders is a form of pollution, of "psychosocial pollution." It is less tangible than material pollution but not less real, and perhaps because of its seeming intangibility even more of a threat to society.

And this concerns only a part of what we have been talking about. Your overall subject of interest with this volume of interviews is, according to its title, the future of life. We have talked about a variety of consequences of modern developments in our society that, though they may be beneficial when seen in short-term perspective, carry grave long-term risks, and even already occurring damage, with them. Without elaborating, which would be impossible within the frame of this interview, I can only repeat what I have said twice in this discussion: unless we change our life styles drastically, the health, physical and mental, and the very survival of our species is in great danger. And I would also like to repeat that change is nowadays so rapid, and we have already gone so far on the road towards worldwide illness and even death that, if we allow matters to go on even for ten years, we will inflict terrible damage on our own children and grandchildren, on people already with us, people whom we know and love. My emphasis is on the words "*if* we allow matters to go on," "*unless* we change our life-styles"; but this change in life-style is terribly urgent, and it can only come about by convincing the masses of our fellow citizens that our position is indeed extremely dangerous; that this situation is of our own making; that science gives us the means to understand and to remedy our illness; and that what is lacking is not the insight but the will and the foresight. We biologists have the duty to make this as clear as we can; and also to stress the analogy that I drew with the long-term policy that is required and *that we do apply* without grumbling too much when we plant trees and when we pay life insurance premiums.

As I said, nothing to do with Christianity; I am speaking as a scientist. There are many good reasons to believe that the group of extended families has been the normal social environment for children for a couple of million years.

The modern "anonymous" mass society, with all that this entails, is far from ideal for a healthy mental development of children. Fortunately, "human nature" forces us to create such "in-groups" within the framework of even large cities. The less extreme type of Israeli Kibbutz, were parents once more play an important part, comes as near to what I consider the optimal situation as seems possible.

M.S.   Or the Asian family, with three generations under the same roof?

N.T.   That is also a good pattern. It is only very recently that families have ceased to be extensive. Personally, when I was growing up, I almost lived in an extended family, since my grandparents on both sides lived close to my parent's home. As a child I could walk from the one to the other house.

JEAN BERNARD
A Distinguished Professor of Medicine

JEAN BERNARD IS DIRECTOR of the Institute of Research on Leukemias and Blood Diseases of the University of Paris, and it was here that the first long remissions of acute leukemias were obtained.

A hematologist of worldwide reputation, a clinician, biologist, and elegant and prolific writer, Jean Bernard has dominated French medicine during recent decades with the tranquil, ironic, and somewhat distant authority of a man who has spent his life in a hospital and university setting, rather like a sultan's residence, and has enjoyed and exhausted all honors, including membership in the academies (the Académie Française and others).

An unquestioned adviser to all governments, at once eloquent and discreet, Jean Bernard would be the archetype of the distinguished professor in the French mandarin tradition—he who makes his rounds, sometimes at a run, sometimes at a slow and majestic pace, followed by a contingent of assistants and students—if he had not retained his freshness, curiosity, and innovator's ardor. All of which make him, at seventy, a surprisingly young man, heedful of our time—of its passions and miseries.

The lineage of Claude Bernard and Laennec has not yet died out.

M.S. Between a miracle cure for cancer and a contemporary version of the elixir of youth, a whole spectrum of drugs and methods outlines the contours of future medical practice. What do you think we can reasonably expect to see happen and in what ways might this be at variance with our vision of utopia?

J.B. One can dream of an imaginary country where an ideal government rules over a happy people. Utopia is a political or social vision that does not take reality into account. André Gide wrote somewhere that "the reality of tomorrow is the utopia of yesterday or today." I'm not sure that that formula is true for medicine. More precisely, one can make the following comments. Of all the methods or techniques that have changed the fate of the sick person, from antibiotics to the Rhesus factor, from insulin to heart surgery, we cannot cite a single one that was not the result of rigorous well-informed research benefiting from the methodology that has come down to us from Claude Bernard. Our hoped-for progress will come from a better knowledge of observed phenomena and from a more profound knowledge of the relationships between structures and functions. Progress requires the rejection of dogma and a respect for method. Finally, we can, I would say, reasonably hope to have, in the first half of the twenty-first century: 1) drugs against cancer based not on the destruction of cancerous cells but on a knowledge of the very mechanisms that govern cancerogenesis; 2) psychotropic drugs based on a knowledge of the physicochemical properties of neurons.

Never has a discovery been made due to nonorthodox methods. All discoveries involve three factors: a profound knowledge of the subject; a mind strictly applied to the subject, concentrating on it only; and, finally, luck. Luck was with Fleming when he discovered penicillin, and with Newton also: If the apple hadn't fallen, we'd still be waiting for the laws of gravity.

But luck is very different from magic.

M.S. You have been a member of official groups that, within the framework of the Massé Plan, tried to foresee and plan the future of the country. Did a futurology of health care seem possible to you then—as was apparently the case for the steel and textile industries?

J.B. The group was called "1985." Along with a group of friends, I compiled the "Health" section of the book that we published at the time. We were not asked for a futurology of scientific research, which is very difficult to predict and, in my opinion, misleading when attempted. But we were certainly asked for a futurology of health care. At the time, we were part of a task force that had been formed by General de Gaulle and Georges Pompidou to make a twenty-year forecast, not the four-year predictions, as earlier required by the framework of the plan. I'm proud to say that the section on health care was the only one in which there was virtually no error. For all the rest—steel, textiles, energy—the forecasts proved to be extremely deceptive. For example, the oil crisis was not predicted, although there were oil men in the group.

On the other hand, we clearly indicated the discrepancy—which was not all that obvious at the time—between health expenses and the GNP, and indicated that for the period beginning in 1985 (and we are getting there) this gap would be a major problem. Also, we specified that, of all the solutions proposed, the only reasonable one was progress in preventive medicine.

We advised the government to increase research in that field. It would be interesting, now, to compare what was said then with present reality. We no doubt came up with lots of stupidities and a few truths; but, when it comes to forecasting, a little truth suffices.

M.S. Will the population explosion make people more aggressive?

J.B. I'm not certain that they'll be more aggressive; but it is unquestionable that the demographic phenomenon is going to dominate the twenty-first century. What is foreseeable is a mixing of populations—due to high birth rates in some and low birth rates in others, leading necessarily to migrations (probably peaceful) and, consequently, to the mixing of different groups.

It is difficult to predict whether these new people will be more aggressive or less aggressive; but they will probably be better, because all modern biology testifies to the advantages of hybridization. It's an idea that we believe in very much. In fact, hybridization constitutes one of the great discoveries of our time, thanks first of all to the Englishman, Allison, and then to Jacques Ruffié, who also played an important role in its formulation.

It is clear that the product of hybridization, or racial mixture, is better off than the nonhybrid individual. The most telling example has to do with sickle-cell anemia, a disorder to which only blacks are subject.

This affliction is due, as we have known for some twenty years, to an anomaly of the hemoglobin cells, which, as its English name, suggests, assumes an S shape.

Like all hemoglobin disorders, this one is recessive. In other words, if we give the name A to normal hemoglobin, we can distinguish three types of subjects: the normal AA individuals, the SS individuals afflicted with this anemia, and the heterozygous AS individuals who are carriers of the defect but are not themselves sick. Allison observed that this third group survived in large numbers with respect to the other two groups, and that an SS group still existed even though, according to the Darwinian laws, it should have disappeared long ago. The explanation of these phenomena resides in the discovery that S hemoglobin protects against malaria. Everything happens as if the malaria parasite "broke its teeth" on hemoglobin that is more rigid than normal. Since malaria is a very ancient disease, the result has been that throughout the ages a polymorphous equilibrium has been maintained in which the A and S genes are simultaneously present. The AS heterozygous subjects are greatly favored, both with respect to the homozygous AA individuals, who are more afflicted with malaria, and with respect to the homozygous SS individuals, who usually die of the anemia.

Some people think that everything began with malaria and that the S hemoglobin appeared later as a mutation in man's defense against that disease. Ruffié and his co-workers reported the following observation made in French Guyana: The Indians of Guyana were veritably decimated by malaria. Then came slaves deported from Africa who were carriers of S hemoglobin. Because of the mixture of the two populations, more and more heterozygous AS individuals resisted malaria. Ruffié ascribed the considerable drop in malaria in Guyana to the introduction of the S hemoglobin.

The idea that "pure" races are superior is a pure illusion, even if it is the main argument of racist theories that argue for it. Our scientific knowledge has proven incontestably the advantages of hybrids and people of mixed race. That's why I have no fear of population mixtures—far from it. The Gallo-Roman mixture, like American society, is a successful model of hybridization.

M.S. Is genetic engineering the promise of a Golden Age? Or does it presage an apocalypse?

J.B. Neither one. Genetic engineering is a very promising methodology that in the near future should, without ushering in a Golden Age, facilitate the preparation of drugs, hormones, and vaccines, and thereby diminish the unhappiness of sick people. At least in the short or middle term. In the long-range view, whether we will have a Golden Age or an apocalypse will depend not upon the doctors but upon the dictators. I believe one must make a sharp distinction between two things. Although we can hope that in the next thirty years one drug out of every two will be manufactured with the help of genetic engineering, I am less optimistic about ultimately passing on bacteria to mammals and man. In that case, we won't escape Big Brother or the Brave New World.

M.S. Do you believe that eventually people will be able to live 120 years?

J.B. Yes, it will be possible to live 120 years, either because we finally discover the genes that determine longevity, or because we learn to create hybrids by crossing the eternal cells—the cancerous cells—with normal cells. That will be desirable if we manage to obtain, starting with indifferentiated cells, the differentiation of the nervous system. I'm referring here to the work done by Etienne Wolff and others, which so far has been carried out only on lower crustaceans—shrimp. In the first place, I believe in the existence of genes that determine longevity, and think this will eventually be confirmed. Once that happens, man will learn how, in the long term, to modify the genetic code. The second key to prolonging life resides in the hybridomes, which result from the fusion of a normal cell with a cancerous cell; the latter confers immortality on the whole. We already know how to create these hybridomes in the laboratory.

M.S. Can euthanasia, under sociopolitical regulation, be integrated into a new morality?

J.B. Everything depends on what you mean by the word "morality." If the word "morality" is taken, as sometimes happens, in a social sense—if it is a question of a kind of social hygiene—the answer is affirmative. Sociopolitical constraints may induce the government in power to consider that euthanasia, the elimination of incurables and infirm people, is good for society. If "morality" is taken in its usual sense, the answer is a matter of indifference. In medicine, the duty of the doctor is governed by two imperatives: the protection of life and the love of one's neighbor. Here we can quote the admirable formula of Paracelsus, carved on his tomb at Salzburg: "All medicine is love." Consequently, no matter what the rules of the state, it's these imperatives that must be honored; we must do what one can in each case, for the better or for the "least bad." I find that euthanasia, about which people talk so much, is the perfect example of a false question—on the moral plane, I mean. Unless we're talking about one of those totalitarian systems of which, unfortunately, only too many abound, I can't see how one can tell a doctor that in cases A & B he or she must do everything to help the patient survive, while in case X the patient will be allowed to die. Each of us, as a doctor, has had incredible experiences in this respect. Family affairs fall rapidly into the realm of street comedy, alone with sordid stories of inheritance.

It is one of the *"servitudes* and *grandeurs"* of medicine to take on the pain of people to the very end, and to try to do the least ill every time. Each situation is a unique case. Any codification in this area is impossible, even dangerous.

M.S. Does man have reason to fear that his free will, his freedom, will be threatened by the new psychotropic drugs? Is it possible, within the framework of a democratic society, to protect oneself against excesses in the use of drugs, the manipulation of the psyche? And if so, how?

J.B. With the psychotropic drugs, yes, the danger is real, in democratic societies as well as others. A highly democratic country, Switzerland, has protected its citizens against goiter by discreetly putting iodine in the salt—without saying so, or almost. Other democratic countries could, with the best intentions in the world and with the weight of scientific reports behind them, even more discreetly put tranquilizers or stimulants into the drinking water, for example. And the dictatorships will do it even more easily, according to whether the political necessity of the moment demands 30 million sheep or 30 million tigers.

But in our democracies we must insist upon widespread public information as a warming to the citizenry. I might add that it is not absolutely clear whether or not substances put into water or food are more dangerous than radios, television sets, and other means of manipulating opinion.

M.S. It's not easy to impose bans and constraints in a free society. That's the other side of the coin.

J.B. There is one answer to your implied question: better educate the public. To try to awaken it, to sensitize it. I know it's not easy.

We should completely revise the educational system and teach the rudiments of modern life to children, preparing them for the problems they are going to encounter in the course of their lives. A boy of fourteen is perfectly capable of understanding what nuclear energy or medication is—the advantages and disadvantages. Then, later, when he becomes an adult, he'll be better able to judge. If the educational system were better—which it is not, probably because of both negligence and pressure from a whole series of interest groups—people would have a greater chance of controlling their lives. They would not have to resort to psychotropic drugs.

And this brings me to the following story, which strikes me as very revealing. During a recent airplane trip that I took, a fellow passenger felt a bit ill. The hostess asked over the loudspeaker whether anyone had a tranquilizer. Everybody rushed over with tablets. I was the only one who didn't have any. The abuse of psychotropic drugs is only part of the general problem of the abuse of medications. In my opinion, the blame for this abuse must be divided among four culprits: the uneducated patient; the permissive doctor; the pharmaceutical firms; and, finally, the government.

M.S. Can the mentally ill really benefit seriously from the use of mind-altering drugs, or *any* manipulation of the states of consciousness, either by psychotropic drugs or by electronic means?

J.B. If you will pardon me, I'm not sure I like the tendentious way you pose that question. Our present shortcomings in psychiatry are due essentially to our ignorance. Today's psychiatry is about on a par with the medicine of yesteryear. It is dominated by discourse, as was French medicine in the nineteenth century. This is not good. In fifty to a hundred years, the chemistry of the brain cell, and its physics, will no longer be a mystery, probably thanks to techniques we can't even imagine now. Medication based on specificity and selectivity will be manufactured. In the beginning people will waver back and forth—as was true for the different kinds of meningitis and endocarditis. Then success will come.

Perhaps this comparison will better illustrate what I have to say about the situation of psychiatry today. A hemophiliac is hit by a stick and dies. The doctors present are divided into two camps. Some of them say, "You see, he died of the blow from the stick." The others say, "No, not at all. Many people are hit by sticks and don't die. He died because of the trouble with coagulation."

The problem with psychiatry is posed in the same terms. An Oedipus complex may perfectly well be due to the existence of the mother, the murdered father . . . but it can also be explained by physical and chemical changes in the neurons. Thus, you will cure Oedipus either by understanding the complex (that's the role of the psychiatrist) or by understanding the chemistry of the brain. That's why I believe it's useless to quarrel: we find the same old opposition between etiology and physiopathology.

As for the electronic means used in the United States—psychiatry via computers, biofeedback—I don't have much faith in them. All that seems premature to me. It so happens that today we understand the chemistry of almost all the cells of the body—the white corpuscles, the cells of the pancreas, etc.—everything but the cells of the brain. We have discovered the transmitters; that is, we have identified the telegraphic wires but not the station. This is a serious hole in our knowledge, but one that in my opinion will be filled, thanks to the remarkable work of people like Guillemin and Changeux.

M.S. What kind of sexuality will we have tomorrow, what with increasing dissociation of the pursuit of sexual pleasure for its own sake and reproduction?

J.B. Two important factors will modify sexuality considerably. I'm thinking of sperm banks and vaccination against pregnancy. The creation of sperm banks is an event of great import. A good number of young Americans have their vas deferens tied off after "depositing" their sperm for freezing in a sperm bank. Thus liberated in their sex life, they can choose when they want to have a child. This dissociation of physical love and reproduction is a new and unique phenomenon in the history of mankind. It offers much food for thought to philosophers and theologians. The second major determinant of future sexuality is the contraceptive vaccination. After several failures, we are not far from success. The first experiments centered on finding a vaccine against the antigens of the sperm. That research was done in India at the Tata Institute, which is the biggest center for biomedical research in Bombay. Having myself worked for a while at that center, I witnessed a rather "picturesque" situation: The Indian researchers had managed to manufacture a vaccine that was so specific, it was active only against the sperm of the husband. The rather droll consequence was that every child born to a vaccinated mother was necessarily the product of adultery.

Another avenue of research that strikes me as promising in the ten or fifteen years to come is vaccination against the pregnancy hormone—the gonadotropic hormone. The idea is to couple the hormone with a well-known vaccine, the tetanus anatoxin. In this way, women can be vaccinated for three to five years.

Contraception is a very old notion. Pliny the Younger recommended to Roman women that they sneeze during ejaculation if they wanted to avoid pregnancy. In spite of everything, I'm struck, however, by the extreme rapidity with which social practices have evolved.

For a long time there was a double standard between the sexes: Society found it perfectly normal that a boy should "sow his wild oats" (as they used to say) but sexual freedom for a girl was frowned upon because of the risk of pregnancy. What we see now is a just reversal of fortune!

M.S. Can you imagine a world where, what with prosthetic devices and transplants, defective organs will be replaced like the parts of an automobile? What ethic must be devise to govern the organization of the organ banks that will be necessary?

J.B. You're touching on Jacques Attali's ideas. In spite of certain rather premature pronouncements, prosthesis and transplants do not represent a definitive evolution within medicine but rather a transition period we must live through before any real progress is made. Medicine, it has been said, either corrects or replaces. It replaces when it cannot correct. When it really progresses, it corrects. It would be difficult to create and label a definitive world of organ banks. There is no problem in perfecting an ad hoc ethic for the rather brief period—some thirty-odd years—of successful prostheses and transplants. I should like in fact to emphasize the importance for research in general of marrow transplants, which have provided us with some very precious information applicable to immunology, the treatment of leukemia, and fertility.

M.S. And yet organ banks pose ethical and political problems.

J.B. Hematologists have the good luck not to be present at the time of mutilation. Actually, they "appropriate," if you will, organs or parts of organs that repair themselves (bone marrow, for example). On the other hand, they do not escape running the risks of anesthesia, although they are slight. By contrast, kidney transplants, where a mutilation is basic to the procedure, pose grave ethical problems, since the donor must be compatible with the recipient.

I remain resolutely opposed to an excess of legislation in this area. Whether it is a matter of transplants, euthanasia, or abortion, only four channels lead to the solution of an ethical problem: the patient himself, if he is capable of making a lucid decision; the patient's family, if it is not too greedy; a committee of experts named by an official authority; and finally—and for me this is the best alternative—a collegium of doctors. Even so, the doctor handling the case bears the full responsibility of the decision. A real difficulty arises here, because young doctors are not prepared to face up to the heavy responsibilities of these very dramatic cases.

M.S. They haven't really even been taught therapeutics. How could they have been taught how to behave when faced with situations of life and death?

J.B. You're right. Not even therapeutics. . . . There has been a lot of discussion about the role of the "councils of order" (*Conseils de l'Ordre*). It's a general question that I won't go into. But in any case those councils have a basic role to play here. They should be "thinking booths" for this type of problem. They should be where the doctor goes to for his information, matching his opinion against those of his colleagues, consulting experts, etc.

M.S. Is it possible that one day immunology will succeed where chemotherapy and surgery have failed?

J.B. Everything depends on how you define the word "immunology." If you take it in the broadest sense, the answer is largely affirmative. One of the principal functions of the individual is the recognition of self, the ability to differentiate between the self and the non-self. The study of the HL-A system—discovered by Jean Dausset, who works at this institute—has shown the unique character of each human being. We have now counted more than 200 million combinations within the HL-A system alone. And we have since discovered that belonging to an HL-A subsystem can carry with it a predisposition to certain diseases. The future belongs more to immunology and the mastering of the mechanisms of cellular regulation than to chemotherapy.

M.S. The management of health care by data processing is viewed by many as the prelude to a more policelike society tomorrow. Do you agree?

J.B. The fact that such a danger exists does not mean that a remedy cannot be found. The risk that the spread of data-processing techniques may lead to social abuses and repression must be pondered and weighed against the advantages of data processing for the patient. In a book published early this year, I visualize the hospital of the future. I believe that in the future we will have hospitals without patients, because the latter will stay home.

We have established in France—as in the United States—"day hospitals" and "home-care units"; but in reality I believe it is all going much further. Actually, we must be clearheaded in our analysis of the situation. We have substituted a new barbarism for the barbarism of the age of Henri IV, when patients afflicted with typhus, the plague, cholera, or smallpox were put side by side in the same bed. Our new brand of barbarism consists in making men and women live in that strange world—inhuman in many respects—we call a hospital. Even if the atmosphere is not hostile, it is experienced as such, not to mention the annoyances the patients are submitted to: catheters implanted . . . and all kinds of equipment.

So I propose and envision a more humane approach for medicine. On the walls of apartments and houses devices registering information concerning the residents will be registered—information that will be transmitted directly to the hospital or other health center. The doctors who run these machines will receive data, discuss them among themselves, just as we do now, and give instructions to the machines concerning treatment.

And I believe that there is good reason to hope that we will then have more humane medical care, because the role of the practitioner (contrary to what some people say) will be expanded. It will be the doctor who gives orders to the machine, which of course cannot replace his competence. And, of equal importance, that doctor will finally have some time—time to visit a patient at her own home, talk with her, see her in her own surroundings. I sincerely believe that we will then have gained much, both economically and morally.

Medicine has fifty difficult years ahead of it. But if it rounds the cape, it will enter a new and very promising era.

I'm well aware of the dangers of data processing, as I am of those of any innovation in any field. Social and political adaptation to technical innovation is not always easy. It's an eternal problem. It's the old story of the gap between wisdom and technique. Man's knowledge has increased prodigiously in science and technology. If you compare Einstein and Archimedes, Einstein was better than Archimedes. But if you compare Plato to Sartre or to Bergson, no one can say they are more "advanced" than Plato. Wisdom has a tendency to stagnate, while technical progress advances; and the gap formed between the two is dangerous because it leads to the death of any given species. The Diplodocus and other reptiles of the Cretaceous period disappeared because of the disproportion between their little brains and their enormous bodies. When I said this to my friend Ruffié, he replied: "Yes, but that can be straightened out in ten or twenty thousand years." As a specialist in physical anthropology, he is an optimist. He believes that it's a simple delay, and that in the next twenty or thirty thousand years wisdom will have caught up with technology. But no one knows for sure, and many catastrophes can occur in the meantime.

M.S. People hope for so many miracles from the "new biology" that some see it as the answer not only to our therapeutic problems but to the nutritional, energy, industrial, and other needs of tomorrow. Do you believe they're right?

J.B. I'm not too sure what the "new biology" is. But in all likelihood the progress being made in biology is going to transform the fate of man and all living beings during the twenty-first century. Biology was at first a handmaiden of medicine. Then it dominated medicine; and now the period of empirical progress we have known will be succeeded by a phase of rational progress. People have not taken sufficient notice of this phenomenon. The admirable progress that medicine has made in the past thirty years has been more or less due to luck. We've talked about Fleming and penicillin, and we could cite other examples. But, for some years now, the real medical discoveries being made have been rational, not empirical in nature. Biology's findings are going to alter our lives in many ways: therapeutics, preventive health care, agriculture, energy, zootechnology, and industry, to cite only a few examples.

The century we have just lived through was dominated by physics; the next one will be a time of great strides in biology.

M.S. At a time when we are witnessing the increasing scarcity not only of oil but of practically all raw materials, these panbiology predications seem awfully optimistic. What do you think?

J.B. That's the basic meaning of the report by Gros, Jacob, and Royer.* I don't know whether or not they are too optimistic, but they think that solutions one would not have hoped for twenty or thirty years ago, solutions for satisfying virtually unlimited needs, will be provided by biology. I'm not competent to judge whether or not that is exactly right.

The solution to the problem of world hunger presupposes that development of biological research in that field will be given a priority. I believe there is a good deal of truth in what the authors of the report write: Great hopes are inspired by the direct fixation of the nitrogen in the air by grains, and the "green" oil extracted from a variety of Euphorbia that might be acclimated.

M.S. How do you see the role of the doctor and of "medical power" in the society of tomorrow? The public sometimes reproaches doctors for their arrogance, their inability or refusal to talk with the patient, presumptuously considered by the medical profession as incapable of managing his own health.

J.B. Contrary to popular belief, this attitude on the part of doc-

*Sciences de la Vie et Société, a report to the President of the Republic. (La Documentation Française, 1979.)

tors is beginning to subside as progress in therapeutics continues. Somebody once wrote that doctors were conceited in reverse proportion to their abilities. Those "ivory tower" doctors spouting the most Greek and Latin phrases—in short, the pedants—were actually masking their ignorance. But today we should not underestimate the effects of technology on medicine—for this has made it difficult for the doctor to communicate with his patient. To make matters worse, medical students are not taught, in the course of their studies, the basics of psychology, of pedagogy, etc., that are indispensable to a real dialogue with patients. Some young doctors do it instinctively, but this is not the case with the great majority. Personally, I devote the first hour of my working day to families. In the course of my experience I have become more and more convinced of the importance of discussing a given case with the patient and even his family, making as much information as possible available to them. On the other hand the doctor does not have to switch roles with the patient. For example, it would be quite blameworthy to conceal from a man of thirty with Hodgkin's disease that the treatment might make him sterile. Yet certain informed patients have refused the treatment and have died. Should they have been told the truth?

M.S. That's what I boil the problem of "medical power" down to in the final analysis. Should one tell the patient the truth, or conceal it?

J.B. The problem isn't that simple. It's still the type of situation for which it's impossible to set forth rules, but one doesn't lie—unless one is incompetent. When the doctor is sure of positive results, he has no need to lie. Thus, there is an obvious link between the failure of medicine and the lie.

In theory, the best way to reduce lying would be to eliminate failure. In the end, however, we find ourselves, I believe, faced with three kinds of cases. The first is that of a disease for which there is a chance of a cure. Here you owe the patient the total truth. You don't have the right to remove a woman's breast without telling her, "You have cancer of the breast." You don't have the right to make a young man sterile without revealing to him that he has Hodgkin's disease.

In the second case, the end is near: The patient is going to die in two or three days. In my opinion, we owe the patient the whole truth here too, because he may have religious matters or a will to attend to.

In the third case—a very difficult one, and for which I cannot give any strict answers—death is certain but deferred. The most typical example is that of chronic myeloid leukemia, which at present is always fatal in three years on the statistical average; that is, in from two to four years in 100 percent of the cases. If you don't tell the truth, you are doing a bad thing, because you are acting as if *you* are the patient, substituting *yourself* for the patient, who might say: "I'm not going to work anymore. I'm going to let everything go to hell and 'have a ball' because I'm going to die anyway in three years." But, if you tell the truth, you create a dramatic situation. I can cite the cases of two young women who were told the truth abroad. Their lives are completely ruined, although they are in good condition right now; and, after all, it's not impossible that within three years a new and effective drug will be on the market.

Finally, the problem can be summed up as follows. For the disease where there is a percentage of genuine cures, no difficulty. For the disease where death will be immediate, you owe the truth to the patient. For the disease where death is certain within three or four years, I don't know. . . .

One of my friends, who was one of the great biophysicists of our time, came to see me fifteen years ago and said: "I'm going to die within two years at the most, and I've come to ask you to do me some favors. . . ." He told me the following story. He was coughing a little. X-rays revealed the presence of a huge tumor of the mediastinum, a reticulosarcoma.* Knowing that, he engaged in a pathetic ruse. He went to see a friend who was a specialist, and told him: "I've received this X-ray of a friend who lives in the provinces. What do you think of it?"

The specialist fell into the trap and replied unsuspectingly: "It's a reticulosarcoma, and at present there's no treatment for it."

"How much time does he have left?"

"Two years. . . ."

Then the biophysicist came to see me. He was very moved when he told me his story; and he asked me to make certain arrangements at his death for his relatives and his students. He had arranged everything. He lived two admirable and serene years, although he was completely atheistic. Some people think that a religious faith can help one to deal with impending death. But in my opinion, not more than 1 percent of humanity is capable of a stoicism like his. He had great self-control and was of exceptional lucidity. He was right: Too many people think and live as if they were immortal.

M.S. Is there such a thing as preventive medicine that is not coercive?

---

*Malignant proliferation inducing tumors in the bone marrow.

J.B. We must remember that there are several forms of preventive medicine. But I must first mention two past forms of preventive medicine that, frankly, were coercive. First, the quarantines imposed in the fight against leprosy. They yielded remarkable results, but the rigor with which the Middle Ages applied those measures was a prime example of prevention-coercion. And then we have the situation presented by vaccination programs: the conquest of smallpox, the conquest of polio, the conquest of tuberculosis. Those countries, like Great Britain, that have not imposed compulsory vaccination have had occasional epidemics.

Despite their very coercive character, these quarantines and vaccination programs have been very beneficial. Naturally, one may think that the line between coercion and strong persuasion, as in China, is very thin. Systematic "check-up" examinations of entire populations have been used by various countries as a means of disease prevention. I am personally opposed to such compulsory examinations, which are usually of no interest. The preventive medicine of the future will be based on the HL-A system, and other systems not yet discovered.

We are nearing the point of being able to give a precise definition to what we witness as a predisposition to disease. We already have that definition in the case of diabetes and chronic polyarthritis. Knowing that an individual belongs to an HL-A group will make real prevention possible, with the consent of the patient. Specific advice can then be given—advice that the patient will be quite free to follow or ignore.

But the situation will not be as rosy as it may seem at first glance. Because the individual and society's interests will sometimes be divergent, I can't conceive of preventive medicine that will not be a bit coercive in one way or another.

If you don't vaccinate a recalcitrant individual, he could transmit smallpox to an entire population. If force had not been applied, we would not have witnessed the eradication of smallpox in the world. I will qualify a bit what I have just said. Anglo-Saxons have such civic spirit, based on the Puritan tradition, that it is less dangerous to leave them free to determine such questions for themselves than it is for Latins. For example, vaccination against polio, which was not compulsory in the United States, was adopted voluntarily by that population, and there was no catastrophe. BCG, despite all the criticism, is a very good method of protection but is not so universally accepted. And as a result the prevalence of tuberculosis is closely linked to the rate of vaccination. It was one of Mao's strokes of genius to have enlightened communist China in 1949 on the prevention of disease. In large part, the Chinese communists based their political success on that idea. They told Chinese mothers: "Your children are dying of cholera, of plague, of malaria. Listen to us! Follow the Party's orders! Exterminate the flies, mosquitoes, rats, and so on." The entire population got to work. Five years later, most of the epidemics had disappeared. Then the government told them: "You can see how well we have succeeded! Follow us in everything else."

And that is how political coercion was later introduced. The Chinese leaders were very clever.

M.S.   Lenin had already said that the formula for revolution was the Soviets plus electricity. . . .

J.B.   And for Mao, it was the communes plus vaccination and the extermination of flies. The campaigns' effect on the minds of the people was prodigious—to the point where I myself got caught up in it. On one occasion when I was in China I was present at a meeting of a hygiene committee. Suddenly, a fly came into the room. Everything stopped, and someone went to kill the fly. When I came back from that six-week stay, I went to Touraine to take a few days' rest. To my wife's great surprise, I seemed very nervous for no apparent reason. She asked about it, and I showed her a fly that had just come into the room. That was the explanation. I had acquired "Chinese" reflexes and had lost the habit of not seeing flies.